LINKING THE NATURAL ENVIRONMENT AND THE ECONOMY:
ESSAYS FROM THE ECO-ECO GROUP

Ecology, Economy & Environment

VOLUME 1

Linking the Natural Environment and the Economy: Essays from the Eco-Eco Group

edited by

Carl Folke

Beijer International Institute of Ecological Economics,
The Royal Swedish Academy of Sciences
and Department of Systems Ecology,
Stockholm University,
Stockholm, Sweden

and

Tomas Kåberger

Physical Resource Theory Group,
Chalmers University of Technology,
Gothenburg, Sweden

KLUWER ACADEMIC PUBLISHERS
DORDRECHT / BOSTON / LONDON

ISBN 0-7923-1227-9

Published by Kluwer Academic Publishers,
P.O. Box 17, 3300 AA Dordrecht, The Netherlands.

Kluwer Academic Publishers incorporates
the publishing programmes of
D. Reidel, Martinus Nijhoff, Dr W. Junk and MTP Press.

Sold and distributed in the U.S.A. and Canada
by Kluwer Academic Publishers,
101 Philip Drive, Norwell, MA 02061, U.S.A.

In all other countries, sold and distributed
by Kluwer Academic Publishers Group,
P.O. Box 322, 3300 AH Dordrecht, The Netherlands.

02-1292-250-ts

Printed on acid-free paper

First published 1991

Reprinted 1992

Printed in the Netherlands

CONTENTS:

PART III
ENVIRONMENT AND ECONOMY IN DEVELOPING COUNTRIES

PART IV
SYNTHESIS

Preface

Severe environmental degradation and global environmental problems have brought the issue of sustainable development into the forefront throughout the world. The Brundtland Report and the recent United Nations Conference on Environment and Development, in Rio de Janeiro, have contributed to a growing awareness and a concensus of opinion that it is no longer possible to take environmental goods and services for granted, instead they have to be integrated in decision-making at all levels in society. Environmental issue simply can not be treated as external, since the precondition for human welfare is healthy environments and functional ecosystems.

There is a pervasive need to understand the interactions between the natural environment and human systems, and there is increasing research on the interface between ecology and economy. Still, among scientists there are, however, considerable differences in perspectives, in the approaches to environmental problems, and on what actions are required to redirect society towards sustainability. With the purpose of bridging and increasing the understanding of the different perspectives among ecologists and economists and stimulating communication and cooperation concerning environmental issues, the Eco-Eco Group was founded in 1984. The initiative in the formation of the group was taken by Johan Åshuvud who, tragically, died in a car accident in 1988. To honour the memory of a skillful colleague and a dear friend we decided to write this book.

This is the second edition of the book and, as the first, it consists of three parts. Part one presents various perspectives on ecology and economy linkages. Part two consists of empirical analyses of the role and value of the natural environment for economic activities, and it provides examples from regional land use, agriculture, wetlands, and coastal and marine ecosystems. Part three deals with human impacts on the natural environment in developing countries, from a local to an international level. Finally, a synthesis outlines the major perspectives in ecology and economics dealing with the interaction between humans and nature, and places the chapters of the book within these perspectives. In this second edition chapter 12 has been rewritten and chapter 14 slightly extended.

It is our hope that the essays from the Eco-Eco Group will contribute to increased communication and reciprocal understanding and openness towards various approaches aimed at linking the natural environment and the economy, for the mutual benefit of all.

Stockholm, October 1992

Carl Folke Tomas Kåberger

Acknowledgements

This book has contributed from the assistance of several persons. First of all we would like to thank Tensie Whelan, Johan's wife, for her interest and help during the progress of the book. All papers have been reviewed and we are grateful to John Bowers, Cutler Cleveland, Robert Costanza, Pierre Crosson, Herman Daly, John Dixon, Karl-Erik Eriksson, Frank Golley, Rudolf de Groot, Charles Hall, Robert Herendeen, Bengt-Owe Jansson, Nils Kautsky, Bengt Månsson, Per-Olof Nilsson, Richard Norgaard, David Pearce, Ronny Petterson, David Pimentel, David Rapport, Rafal Serafin, Thomas Sterner, Lori Ann Thrupp, James Zucchetto and Thomasz Zylicz for their constructive comments and for kindly helping us to improve the papers. A special thank goes to Ulrik Kautsky whose interest resulted in a generous financial grant from the Swedish International Development Authority (SIDA) for distribution of the book. The most welcomed grant from SIDA made us indeed very happy, and is of course gratefully acknowledged. Funding for the project was provided through grants from the Swedish Forestry and Agricultural Research Council (SJFR), and the Swedish Council for Planning and Coordination of Research (FRN). We do appreciate their support to the work with this book and to the research field of integrating ecology and economy. The english was corrected by Chris Wilson and Tensie Whelan. Many thanks also to Olimpia Garcia-Flores and Meta Kågesson who assisted with typing.

A Dedication to Johan Åshuvud

Johan Åshuvud, our dear friend, skilful colleague and the founder of the Eco-Eco group, died tragically in a car accident in May 1988. The sad message of Johan's death reach us at an international conference on sustainability issues and the future of our globe, with participants from both the natural and social sciences. Johan was deeply engaged in such issues, and despite his young age (born February 26, 1961) he had contributed substantially to an increased understanding of the interdependencies between the natural environments and our socio-economic systems.

Already in 1979, as a fellow at the Lester B. Pearson College in Canada, his enthusiastic promotion and dedicated work while at the college, resulted in the creation of one of the first marine reserves in the world, The Race Rocks Ecological Reserve, Victoria, British Columbia. He took his master's degree in economics at the Stockholm School of Economics, and continued with Ph.D. studies in natural resource- and environmental economics. He also studied at the International Institute for Applied Systems Analysis (IIASA) in Vienna.

Johan realized at an early stage the importance of linking the natural environment and the economy and worked enthusiastically to speed up this process. His concern about growing environmental threats, problems with food scarcity in the third world, and an ambition to try to do something about it made him feel he did not have the time to wait for abstract scientific solutions. He wanted to deal with the problems in the field, in reality. For a while he studied the environmental and economic consequences of the construction of a paper mill in Tanzania. He spent his last two years with his wife Tensie in Costa Rica where he worked as Regional Project Representative in Central America for the International Union for Conservation of Nature (IUCN).

In a tribute to Johan, Mark Halle at the IUCN wrote the following "There are few instances in the history of IUCN´s field operations where such a significant programme of activities, and such an extensive network of new partnerships, has developed in so brief a time. Most of the impetus for the Central America programme´s rapid yet systematic growth, was provided by Johan Åshuvud. Johan built for IUCN a position of respect and trust in the Central American community that is truly remarkable. How did he do it? The answer lies in characteristics of Johan´s which, in many ways, mirror the profile that IUCN would like to be known by as an organization. Johan had vision, dynamism, and a sense of commitment to the goals of IUCN in Central America, which guaranteed his effectiveness. But, more than this, he had the character, honesty and openess to be a real "integrator" - of project components, of ideas, of people. Johan was an open and cooperative person. He had a basic trust in people and in the positive nature of their intentions, which in turn instilled in them a sense of confidence and a desire to cooperate. Johan also enjoyed an exceptionally sunny and personable character. I have never before encountered a person like him whom everyone liked on first contact, and for whom everyone´s respect continued to grow on further acquaintance. He made people want to work with him and inspired them to work together. Johan has left behind a solid programme of activities (see Åshuvud, Chapter 13), a constructive place for IUCN in Central American conservation, and a heightened interest and commitment to conservation as an essential component of social and economic well-being. His memory continues to stimulate cooperation and new initiatives in the region. The challenge is with IUCN to ensure that we do not let down the ideals he represented, and we intend to meet that challenge."

We miss Johan, and feel a deep grief for having lost a warm, considerate, humble and faithful friend. What Johan gave will live with us forever as a driving force in our continuing work.

Carl Folke and Tomas Kåberger

on behalf of the members of the Eco-Eco Group.

Part I

PERSPECTIVES
ON
ENVIRONMENT AND ECONOMY

Linking the Natural Environment and the Economy;
Essays from the Eco-Eco Group,
Carl Folke and Tomas Kåberger (editors)
Second Edition.
1992. Kluwer Academic Publishers

CHAPTER 1

The Contextual Features of the Economy-Ecology Dialogue

by

Uno Svedin

Swedish Council for Planning and Coordination of Research
Box 6710, S-113 85 Stockholm, Sweden

When the connection between economy and ecology is discussed this is often done, either as a question of establishing a link between two separate modes of knowledge, or as a definition of the overlap of two partially similar domains of understanding. In fact none of these two models provides sufficient explanation power to allow for a picture which is broad enough. First of all we are not talking about two homogeneous domains of understanding. There are several "economies" as well as "ecologies." This holds true both with regard to focus of interest, for example along the micro-macro dimension, as well as for more basic assumptions underlying the various theories such as views of evolution, the role of information etc. In addition, the connection between ecologies and economies as an overlapping pluralistic area of intersecting connections is highly dependent on the contextual framework. This context has many facets, not the least a theory of science context dealing with topics such as what is considered a scientific issue, and how it should be treated. There is also a sociology of knowledge related to the internal value structures of the various disciplinary constituencies which are involved. This framework could be generalized to the overall social context within which basic value issues and their socio-economic and cultural impacts need to be addressed. In all these cases the concepts of the economy-ecology intersect, which for example could be "nature," "the environment" or "natural resources," are in most cases in themselves contextually dependent entities.

1. INTRODUCTION

The interface between ecology and economy has in recent years been drawn more and more into the foreground as one of the key domains for a discussion on environmental problems at large. Why is this so?

A simple answer could be that the environmental issues in many highly industrialized countries during the 1970's and early 1980's increased politically in importance, but were still viewed as sector concerns. Towards the end of the eighties such issues, it seems, have now emerged as overriding matters of public policy on a par with national security and the economic health of nations.

This in turn has made it necessary to address the question of how to reconcile societal economic activity with the environmental restrictions which are seen as necessary requirements for a healthy society in the long term. This need has in the policy arena been well articulated by the Brundtland Commission in its report published in 1987 (WCED, 1987).

The emphasis on a "sustainable development" combining economic growth and environmental concern sets the stage. The important task is to outline a vision in which earlier conservation aims are no longer necessarily seen to contradict development goals. Rather a healthy economic growth, to some extent of a new type, is seen as a prerequisite for finding the means for a constructive policy with regard to the environmental threats we are all facing. In this way, at the policy level and as an expression of a thought paradigm on how to view the path which is needed, "economy" and "ecology" have been connected much more strongly and more intricately than before, as compared to when they were seen more or less as two antipodes.

However, what the Brundtland Commission has stated, partially as a statement of faith and as direction for future work, is in an operational and analytical sense very incomplete. We are faced with a tremendous challenge to fill in the "missing bits and pieces" in order to get the desired sustainability vision to become a precise blueprint. Yet we do not even know if the vague blueprint is operational at all, which indeed is an empirical question for the future (Mykletun, 1990). And it is here again that the emphasis on the interface between ecology and economy emerges, now in analytical terms, as a very important hot spot. All the ambiguities which are embedded in the "sustainability" package of today have to be cleared up exactly here.

This is not an easy task since many of the presumptions and underlying ideas are different in the two spheres of understanding. In fact, we are not only talking about two distinct academic realms of competence, denoted "economics" and "ecology" respectively. Indeed these labels have to signify wider intellectual territories than just the content covered by present-days academic departments assigned these names at the universities. The interface between economics and ecology should rather be seen as an indication of an area of inquiry that many branches of knowledge have tried to address for a long time. It is the realm in which Man's dealings with Nature are discussed within a societal perspective. Especially geographers, but also others, have dealt with the problems of this "hot spot". In the further elaboration of the problems of this interface we thus have to take into account many other traditions than just "economics" and "ecology".

In the eighties an analytical *diversification* has taken place, in parallel with the increased *political* emphasis that has occurred with regard to environmental issues. Today, at the outset of the nineties, the importance of the social sciences and the humanities within the environmental field has at last received more and more recognition. Thus over maybe two decades the earlier natural science dominance has been complemented more and more by contributions from the social sciences and the humanities (Svedin & Heurling, 1988). And there is more to come. In this broader sense, the interface between "economics" and "ecology" is coming into the forefront as a basis for reflection about key issues of an environmental character, though by no means the only one.

Our common work towards knowledge of these processes is a sort of risk insurance against incomplete understanding of a complex system and its partially hidden and unknown feedbacks.

Today it is not possible to know, at least in detail and with certainty, what happens to multi-stressed natural systems when we apply a human impact pressure of varying degrees on them. In many instances it is even very difficult to disentangle what are man-made causes from what are more natural ones. Results from more work on "environmental history" in a long term perspective could in the future contribute to a better understanding in this respect.

But also the socio-economic system with its feedbacks and sometimes strange and surprising behaviour vis a vis the environment needs to be very much more understood in all its intricate details. This is valid especially for issues related to values, human behaviour, and the institutional forms, in a broad sense, through which society has chosen to realize its aspirations.

Having said this you could then ask: "what are the basic problems that must be faced in understanding this "interface" between natural science and social science with the task of providing a better penetration of environmental problems?" A central theme in the following presentation is the underlying differences in *concepts* relating to the issues, varying *perceptions of causal relations*, but also of how *problems should be considered in general*. The mental models used, whether they are formalized or not, have to be scrutinized. Often they have a cultural bias. Specifically it will be argued that the *contextual nature of concepts, ideas about relations* and the *societal embedding* of the problems are at the heart of both the problems and the solutions.

2. CONTEXTUAL FEATURES OF KEY CONCEPTS

What then is a contextual nature of things? It has to do with the "*embeddedness*" in a broader framework of understanding than just the object of scrutiny itself. The concepts for example come *together with* a background of connotations of ideas and perceptual "packages". Why and how could this type of understanding be of any use in understanding the issues of the interface between economy and ecology? Let us, for the sake of example, discuss the conceptual nature of the concept of a "natural resource", but also because it is an important element of understanding in itself. Later we shall move to the concept of sustainability and from these examples at the level of concepts carry the discussion on to the levels of views of causalities and finally to the contextual features of the cultural and social embedding of our understanding.

2.1 The Contextual Features of the Concept "Natural Resources"

The concept of "Natural Resources" is of high importance for us when discussing the ecology-economy interfaces. The reason for this is that the concept covers many features of the debate. It also has meanings at the conceptual level, both from the side of economics and from the side of ecology. Both "sides" seem to understand what is meant by a "natural resource". But is it really the same thing that they are understanding?

The word Natural Resources has its problems especially due to its common use in almost daily conversation. We all think we know what we mean by natural resources at a practical level. The forest is cut down and turned into pulp and paper. So the forest is interpreted as being a natural resource, i.e. a resource perceived to be waiting "out there" in Nature. In a similar way iron ore is found in various places in Nature and could be used, if so decided, to extract iron in order to manufacture the metal. Thus it is regarded as being a natural resource. Even clean air is sometimes referred to as a natural resource in an analogous manner.

In a closer scrutiny, and by analyzing a number of real cases, the concept is related to a set of problems which calls for various types of specifications: a natural resource under which conditions?; for whom?; assuming which set of assumptions? In short: the concept is sensitive to the context in which it is used and cannot with any amount of precision be used without specifying the context.

What then are the problems? A "natural resource" cannot be equated to just any material phenomenon in a simpleminded fashion, i.e. be limited just to what exists "out there in Nature". A natural resource gains at least parts of its resource features through human intervention in various ways; by knowing it is there, by (potentially) knowing how it must be used technically, and by arranging its links to some societal need.

Two of the problems are these:

- there is a dynamic time factor to the concept,
  i.e. "potentiality" is an inbuilt problem

- a natural resource is not a resource for everyone;
  it is actor selective.

Let us briefly discuss some of these problems. First of all you could ask yourself whether or not a "resource" also has to be something related to Man. Does the function always have to be related to a human goal? In economics the goal function always has a human focus. In an ecological perspective this is not necessarily the case. Due to ideological preferences, the argumentation could either emphasize an ultimate "usefulness to Man perspective", or a view trying to refrain from the "human species-ism bias" emphasizing more equal rights for living organisms. At least in this perspective it would be possible to discuss ecological functions as such without necessarily referring to their direct uses in a societal human context.

However, "material resources" normally relate to phenomena which are seen to be, or which potentially could develop into, resources for Man. As such this perspective is closer to the way of thinking in economics, though existence values of Nature are also sometimes referred to in this realm. In turn the ecological perspective emphasizes "life support systems". Through the chains of ecological connections, some known and others less well known, there is a positive use for Man of these "natural resources". However, an ecological science perspective is not seldom connected by a certain individual with a normative position very much caring for the preservation of the "natural" systems. It has to be stressed, however, that such a normative stance in no way necessarily follows from an ecological science discourse.

However, using the word "natural" introduces the notion of "Nature", which is very multifaceted. The ambiguous character of the word "Nature" explains something about its affective power. At the analytical level it calls for specifying the circumstances, in which it is to be used, as well as the mode in which the interpretation has to be made, i.e. in both cases its context.

In a linguistic analysis based on the present-day use of language, Jens Allwood (Allwood et al., 1981) classifies the use of the word into the following main categories:

- Nature as totality, "reality"
- Nature as that Part of the Environment which is Unaffected by Man
- Nature as the non-artificial
- Nature as essence
- Nature as harmony, and as lack of disturbance and strain
- Nature as that which is expected, explicable or evident.

There is an internal tension between the two parts of the term "natural resources". The problem is at which time and under which circumstances the phenomena in the unaffected part of the world could *turn* to become a resource for someone and for a certain purpose. Have the phenomena the inherent property of being resources or is something "extra" needed to turn bare natural phenomena into resources. Here it is argued that there is indeed a key element of human interference in order for any material phenomenon to *become* a resource for Man.

This statement borders on a normative position. The arguments about where the source ultimately is to be found for the usefulness - or resourcefulness - of phenomena can be traced back at least to Medieval times. As a reaction against too Aristotelian a view emphasizing Man's role as a passive receiver of impressions, e.g. Cusanus stressed in the 15th century Man's creative role. Knowledge is not to be seen as a reproduction of reality. It is also a result of Man's own active endeavours. In this way resourcefulness enters at the conflux between enabling factors in Nature and Man's active interference with it.

Scientific argumentation normally tries to stay away from too value-laden normative entry positions. However, it must be argued that at a deeper level already the choice of object of interest, the way of narrowing the scope in order to focus on certain features, the means by which this is done, as well as the historical development in the respective sciences do imply strong connotations of an ideological kind, which in a broad sense could be seen as a sort of socially agreed upon set of notions and world views.

In this context economists normally address the issue of a "natural resource" with quite a "social" connotation, i.e. stressing the importance of the *human* element in the creation of a resource. Also the actor dependent side is more clearly seen in a social science context. This is valid also for the understanding of the importance of the social situation for the "resource" feature to emerge, including the understanding that technological capacities vary with time and thereby influence the resource element.

Ecologists rather tend to emphasize the natural functions and services "as such". The resource capacity is not actor dependent, nor is the changing social context a problem in defining resourcefulness as the "Nature" functions generate their capacities independently of Man. Thus a physical accounting measurement system makes sense in such a perspective, providing measurements of entities "our there in Nature" as the basis for a fair mapping of where "resources are to be found".

2.2 The Contextual Feature of "Sustainable Development"

There is in our time an uneasiness among politicians, scientists and citizens alike, that the biological foundations upon which all life depends are being heavily eroded. The concern takes many shapes, e.g. the concern about losing biodiversity of species, the chemicalization of toxic elements into the environment to an extent which was never there before, or of man-induced climate change, thereby threatening certain forms of biological life in certain places (Clark & Munn, 1986).

In this situation many have recognized the need for a "sustained" performance of the natural world. This has been linked conceptually to the possibility in the social sphere of a continued development, maybe in a different form than the present one (for a survey of "a gallery of definitions" of "sustainable development" see for example Pearce et al., 1989).

But it has to be argued that this "framing concept" of our time is indeed contextual in character and thus cannot be treated as an absolute frame. Instead we have to ask ourselves questions like; Sustainable, how? - Sustainable, for what purpose? - Sustainable, in which sense? - Sustainable for whom? - Sustainable under which conditions? We may even have to define the actor who makes the statement as to what is sustainable and what is not.

So what is sustainability (Svedin, 1988)? Or to rephrase the question: What do we mean when we say something is being sustained? Is it something to be kept constant or is it the capacity of something to have some endurance over a considerable time? It is reasonable to adopt the latter approach, still realizing that at some level, some indicator has to signal constancy of behaviour. The means by which this "constancy" is achieved could either be *static or dynamic*. In the static case, we are considering some prevention of disrupting external factors having an effect on the "sustained function". In the dynamic case in contrast we are discussing the maintenance of the balance of supporting forces which among themselves could change in composition as well as in strength. The important point is that a dynamic equilibrium is maintained irrespective of the means by which it is maintained.

If we agree that sustainability has this dynamic feature, we still have to address the issue of *what it is* that should be sustained. Is it

- the *specific material expression* of something (e.g. a specifically localized grove of trees)?
- the *specific type of material expression* (e.g. a certain type of forest at a certain place with a specific age distribution, species variety, quality of trees, etc)?
- the *capacity to generate a specific feature* (e.g. the capacity to grow pine trees)
- the more *general capacity to generate any biological product* (not necessarily specific species of trees but more generally "biomass", e.g. in a toxic environment).

Thus, we have to move from a *specific material expression* to more *general functions*. But at the same time we also have to acknowledge the spatial expressions of this. Within which geographical domain are we operating? - Are we considering homogeneous or patch-wise types of distributions?

In addition, the spatial dimension has to be combined with a temporal one. For how long should the entity be sustained? The answer probably has to be "a considerable time duration". But this indeed is a very relative type of statement; "considerable time" with regard to what type of measure of time spans? In addition you could have a long term sustainability of an aggregate over a wide geographical territory without keeping any part of it very sustained at the micro level. Or at least the micro level geographical unit could expose quite vivid temporal cycles of performance, as is the case for forest fire prone regions or for that matter of "slash and burn" agriculture.

So you have to ask at which "*systems level*" something has to be sustained, in addition to the issue we raised earlier regarding what types of *features* should be kept, and by whom and for what purpose. But the sustained feature does not need to be limited to the realm of performances of an ecosystem, e.g. in terms of the biomass production capacity. It could equally well address a certain level of material standard for a certain population, or a societal or cultural order of some kind.

We then have to demonstrate the issue of relationships between sustainabilities at one level and those at other connected systems levels. This could in fact go in two directions. There might be a need for a sustained performance at a higher systems level in order for something at a lower level to be sustained.

But we also have to consider the reverse direction from bottom to top. Are certain sustainabilities at lower levels required for sustainabilities at higher levels?

There are many examples of more or less coping types. We have to discuss *coping patterns* or *coping characteristics*. But what are the criteria for the design of "coping patterns"? We may frame our problem in terms of "Veto-rules" or "promotion of sustainability rules". And we have to address the issue in relation to the situational context, i.e. to talk about sustainability promoting situations and sustainability breaking situations, which in turn calls for the introduction of a normative element.

We thus have to introduce the contextual characteristics for keeping something sustainable and what breaks it. What, for example, is the regenerative capacity of a biological species above a certain breaking point of toxic contamination? Or what is the level of lack of basic needs in a population which forces a local destruction of the environment?

We also have to discuss situations which might change characteristics under certain conditions, e.g. which over time flip their behaviour from a sustained to a non-sustained one. The time dimension is in general one of the most important issues with regard to sustainability contexts. Often the intergenerational equity criteria is placed at the core of sustainability definitions.

How then is the contextual character of the sustainability concept influencing the interface between economy and ecology?

First of all the *contextuality* as such has to be recognized. This is probably more easily done in a social science tradition than in a natural science one.

Secondly the *interconnectedness* of the various levels of sustainability must be recognized. Environmental, socio-economic and cultural forms of sustainability mutually enforce each other in complex ways. There are no easy causal links to be identified, rather patterns of causally enabling settings. Often this takes the form of disrupting vicious circles. This web of interlinkages has to be recognized from both sides. The economy side has to encompass more strongly the absolute need for certain environmental functions and making room for it in economic theory. This involves, for example, a limiting of the perception that vast substitutabilities between different forms of assets are in fact possible. The ecological side, on the other hand, has to accept that disruptions in the socio-economical or cultural sphere might not only be a dominant cause of environmental degradation, but also hold the key to designing solutions.

Thirdly the issue of inherent *change* has to be addressed. Ecosystems are not, and have never been, "frozen" entities. They are continuously undergoing change. Here various subgroups within "ecology" as well as "economy" have vastly differing ideological interpretations with strong influence on the ways goals are defined in the environmental arena.

By recognizing and accepting the contextuality of the sustainable development concept and thus opening up for an operational refinement of the use of the concept, some of these problems might be constructively addressed in the future. Or, at least, the differences in value terms could be made more visible by extracting them from the more technical work of becoming more precise in an operative sense.

3. THE CONTEXT OF MODES OF THOUGHT

3.1. Models of Causality

We have now discussed the contextuality of some examples of key concepts. But ways of thinking involve not only concepts and their connotations, but also views on how things are related, i.e. causalities. It also relates to the forms of questions you have in mind when addressing environmental and other issues.

The set of potential issues you are ready to consider as "interesting" and/or "valid" provides a sort of window for what type of thinking is considered to be possible. Related to such an "interest-space" is normally some sort of mental map regarding relationships in general and a sense of where the key points are. This window directs attention and it leaves things outside the illuminated interest area in the form of less prioritized secondary items.

Any scientific approach involves some sort of "torch lighting" procedure. And they differ more or less from each other. The different forms of models - or theory structures - aiming at the mirroring of relations between Nature, Society and Technology could be analyzed in many different ways. Geographer Robert W. Kates distinguishes between three types of theories of the human environment (Kates, 1988):

- one dimensional causality
- partial theories
- interactive theories.

In the first category we find as control ideas the following types; supernatural control, biological determinism, environmental determinism, ecological balance, social dominance and autonomous technology.

The partial theories are often of a dichotomous nature: Nature and Society, Society and Technology; Technology and Nature. Some models of these types are partial "not by their factorial emphasis, but by their limited domain of application." Thus, for example, classical economic production models combine Nature, Society and Technology as land, labour and capital, but limit the explanation to the supply side.

Finally, in the framework of Kates the interactive theories involve various "conceptual triads and quadriads" built around variants of Nature, Society and Technology; place, work and folk (Le Play, 1879); habitat, economy and society (Forde, 1963); man, mind and land (Firey, 1960); population, social environment and technology (Duncan, 1964), social behaviour, technology and resource opportunities (Butzer, 1982). Or conceptually similar: materials, energy and information (Boulding, 1981); environment, subsistence and system (Ellen, 1982); infrastructure, structure and superstructure (Harris, 1979).

To the interactive theories Kates also classifies "accounting systems", such as the H.T. Odum (1971, 1983) descriptions based on exchanges of values of energy; or the Ayres studies (1978), based on materials or Leontief's studies (1977), based on monetary information. Also systems models, like Miller's "Living Systems" (1978), belong to this class, as do the adaptive ecological models of Holling et al. (1973, 1978). Finally the global models of Meadows et al. (1973) are brought to this general category, as holds true of later practitioners in this very field such as Meadows et al. (1982).

What differs between these models is not only the degree to which they have simplified forms of bi-polar causal connections at the core of explanation, or if their causal webs are extended to many forms of simultaneously acting causal conditions. What differs is also to what extent certain classes of explanations are kept outside the set of those considered. What also varies between approaches is the relative strength a particular analyst attaches to the various causal links possible. One example is the varying emphasis given to the population variable with regard to various environmental phenomena ranging from local environmental degradation to global migration streams of environmental refuges, and the final possibility of the collapse of the entire life support system on this earth.

In these cases the realms of "interesting problems" and "connected variables" seems partially different in ecological and economic theory in their application to topics of mutual concern.

3.2 Ecological and Economics Thinking

What then is the framework of ecological thinking? First of all it is normally biological in its attitude. This means that in contrast to physics it has life in its various forms as an object. At the ecosystems level it recognizes an interplay between different agents performing functions, e.g. as transmitters or transformers of material forms and of energy. The relations between these entities are expressed in terms of a systems performance, e.g. the specific chemical transformation of material flows under the influence of, and in changing the quality of, incoming sunlight energy. This system, apart from carrying a life of its own, provides biological "services" in nature, e.g. building up structures of higher structural value than the initial components (e.g. sugar or wood or chlorophyll from basic elements). The services could also be of other kinds, e.g. providing regulatory mechanisms for certain compounds such as carbon-dioxide concentrations in the atmosphere or "cleaning" functions, i.e. separating certain substances from others denoted as "toxic" with regard to certain forms of life at certain levels.

Such services are normally beneficial to Man, who as a biological organism depends on the maintenance of such functions. In a general sense the continuous reproduction of the function of these systems, including the normal periodical changes of these systems, is thus of great importance for Man.

Ecologists have mainly tried to find general theories, to unify patterns and processes in natural ecosystems, and to provide a framework or a backbone for further generalizations. "One characteristic of ecology as a natural science is that it has very few fundamental laws. Rather, an essential ingredient in the theory of evolution is that each individual is genetically unique providing for alertness to an infinite number of possible environmental alternatives. However, the fact that each component is unique in its individual genetic constitution does not necessarily mean that development in the ecosystem is unpredictable. The concept of succession in organisms, defined as r- and K-strategies (exploitation and conservation strategies, respectively), adequately describes two opposite strategies that are successful in different stages of ecosystem development. It provides a useful tool to predict, at any phase of degradation, what development is most plausible. However, natural systems do not always lend themselves to categorization into r- and K-communities. In fact we know of very few K-communities where succession without disturbance has reached a stable and persistent level" (Brinck et al., 1988).

In economic thinking, on the other hand, many of these functions of Nature are taken for granted. Also in cases of economic approaches on which the cost involved in using Nature's services are envisaged, the realm of influence from Man on Nature is considered relatively small taking account of only the most obvious "costs" in this regard. Deeper and more implicitly the assumption about a continuity of Nature's regenerative capacity under societal stress, is taken for granted. The risks of sudden disruptures and irreversible performances are less frequently taken into account. On the other hand, economic theory in general channels a much more realistic view about flexibilities in the adaptive mechanisms in human society, than holds true for ecological theory expanded into the societal domain, e.g. through energy analysis.

The approaches to causalities are thus different in character depending on which perspective is used - an economic or an ecological one. The choice of systems boundaries is different in time and scale, the choice of variables is partially different and the emphasis on various elements differs. This holds true not the least with regard to the role of information in the various systems (Svedin, 1985).

The differences in approaches exemplify the varying analytical contexts within which root causes are emphasized differently and within which the pattern of causalities is thus given different forms.

4. THE CONTEXT OF THE CHANGING VIEWS ON PROBLEMS

As seen from the broader perspective of social science and the humanities (and then not only "economics") there are other facets to be dealt with as well.

4.1 Changed Patterns of Threat

The presently changing perception about the scope of environmental problems also requires entirely new fields of inquiry with regard to the social setting in which these problems occur.

How do the rules apply and how are new reforms to be implemented? What is the role of property rights and legislation both with regard to the national and international scenes? How are international negotiations to be designed around the global commons, so often regarded as free goods? What type of driving economic and institutional forces are "behind" the new threats and which types of economic incentives can be created to improve the environmental situation in different types of economies?

This widened panorama of environmental issues has thus increased the need to address a different set of social issues than was the case before. And this has happened over just the last few years.

4.2 The Global Scope of Effects

This tendency is enhanced even more since causes and effects are not local or semi-local any longer. We no longer talk about a locally polluted lake or air pollution in concentrated areas of urban sprawls only. What is new is the interregional and global scope of effects, as well as the "global" way of perceiving these threats.

Entire energy production systems - such as that in continental Europe - threaten, due to acidification, not only local forests but entire forested regions. Transboundary pollution not only calls for very much.more sophisticated knowledge about production of the pollutants, their dissemination and their plural effects, but also for new designs of interstate negotiations. This also holds true for the case of the pollution of the major rivers, such as the Danube, which pass through several countries. Problems related to energy production and car transport systems (causing amongst other things acidification) or problematic agricultural practices and simplistic sewage management (causing considerable water system damage as in the case of the Baltic) introduce several questions about the design of entire sectors of the economy as well as life style issues.

The earlier kinds of *accidental events* in the environment field were mostly of a local nature even if the effects at that level could be severe enough and even fatal for the unfortunate individuals concerned. The catastrophes of today - such as the Chernobyl nuclear accident - destroy vast areas for a very long time. The genetic effects over generations have so far not been estimated with sufficient certainty.

The *diffuseness*, the *size of the effects* and the *very long time scales of the impacts* point at new serious elements in the environmental field.

In addition, the new types of environmental problems, such as the possible climate changes, have a *probabilistic* character rather than a manifest one, this is also true for the more "classical" environmental impacts we already "see", such as eutrophication of lakes (Svedin & Heurling, 1988).

4.3 A Vast Causal Web

In all these cases, the threats do not stem from occasional hazards outside of Man's doings or control in principle as did once the so worrying volcanic outbreaks, e.g. of Etna. On the contrary, in many cases it is the very heart of our daily routine activities that creates the problems. This calls for new forms of conceptual and analytical "models".

The burning of fossil fuel cannot be avoided easily in our present-day infrastructure, which in many ways provides the basis for our type of development. In some cases, as holds true for the energy production system but also for parts of our transport systems, the changes that need to be made are very intrinsic with regard to the infrastructure. The changes needed thus address our life styles, including the patterns of time use for individuals.

In some cases, however, changes might seem to be a little easier, but still not easy. This would seem to be the case in the reduction of the use of certain chemicals that threaten the ozone layer. But already in this case, which might seem so clearcut and obvious, considerable inertia in the social and institutional system has prevented quick and forceful action. The understanding of the entire system needs to be considerably improved, with regard to e.g. knowing how to exert social pressure, or investigating the institutional forms in which action might be negotiated and channelled.

What we see is the systemic feature of phenomena which are interlinked in a vast causal web. The mapping and understanding of this "web" are the challenge of the environmental modelling activities of the nineties.

We also increasingly have to deal with societal mechanisms, development strategies and their political contexts, national and multinational companies' dynamics and aspirations as well as local interests in their interplay with broader and more non- local interest groups. In many cases the necessary knowledge is not only different from the more direct ecological one, it builds on other research traditions. This changes the forms of the models which can be used, and also highlights the assumptions behind them, i.e. "the world view" implicit in such endeavor.

The new knowledge needed has also the characteristic of being “dangerous knowledge” for those “in power”, since it interferes directly in the high political domain. This challenges the modes in which models are used in the political arena, i.e. the interplay between model content and user needs.

In all these cases different types of models with different research emphasis and originating from different research traditions have to be mobilized.

4.4. Loans of Elements of Thought

These cross-fertilizations from natural science into the social sciences and humanities and also the reverse, might be of key importance in getting a suitable set of model structures in the new situation and provide the perspective on environmental issues we have outlined above. Ecologist C. Holling (1990) has made the remark that there is a great difference between the 1970’s and the potential 1990’s in these regards.

In the seventies the environmental threats were conceived of as local in time and space. Thus change could still be seen to happen gradually and some sort of stability could thus be seen as an achievable goal. At the methodological level this corresponded with interdisciplinary approaches aiming at formal modelling in the systems analysis tradition. The need for large quantities of data to be fed into these models was seen as a bottleneck.

In the nineties, however, environmental threats are perceived as emerging on a very large scale and they are often related to socio-economic turbulent factors. Changes occur abruptly and rare events create structure. It is becoming more and more important to know the crucial difference between what it is not possible to know due to lack of present knowledge, and what is due to genuine indeterminacy. The importance of the instabilities in the socio-economic realm is not the least evident since they often have a spill-over effect on the environment. Thus the capacity “to understand” rather than modelling what you think you already understand comes more into focus.

Thus, also the character of the needed “model images” changes. The role of simple “didactic” mind models (more or less formalized) gains in importance. The need to get a structured way to think about socio-political phenomena in relation to the environment calls for new innovative modes of representation. You could even ask whether this new situation implies a shift in the emphazis from “analysis” to action, or at least to integrated analysis/action/analysis "packages" of impact sequences. The way how to relate to the entirely unplanned surprises in the natural as well as in the social spheres becomes very important (Svedin & Aniansson, 1987).

In a similar way the shift in the qualified public perception from around the early 1980's, no longer so vigorously stressing long term *physical* shortages as the key problem, in accordance with some of the early "Club of Rome" scenarious, e.g by J. Forrester and co-workers, exemplifies the context of changing views of problems. During the 1980's the stress has more and more shifted towards constraints of a more complex kind, often related to the general environmental load and possible disruptions of life-support systems, for example as a consequence of the energy use pattern (Bertelman et al., 1977, 1980; Passet, 1979; Sachs et al., 1981).

The gradually changing image of the problem situation thus provides the contextual framework for the thought process in which various relevant modes of approach are moulded, be they drawn from the "ecological" or from the social/economic entry-points.

5. THE CONTEXT OF THE CULTURAL FRAMEWORK

The evolving view on problems sets the stage in which we develop theoretical tools to come to grips with reality. The perception, i.e. the context which consists of our present-day world view, is, however, local in time and to an even greater extent local in cultural terms. The call for a long-term sustainable development invokes distinctly basic issues about value structures and drastic revision of ethical norms not earlier needed in our time. When the need for an expanded time span into the future is more clearly seen and the insights about the possibilities of future drastic surprises in the man-made, as well as in the natural world, grow the topics about culture and ethics emerge on the list of issues to be dealt with.

Normally Man's relation to Nature is strongly engraved into dominant cultural traits, be they modes of thought, behaviour or life styles. What then could we say about this cultural context and its demands on the economics/ecology interface?

We have to note immediately our strong link to the "Western industrialized world" leaning towards some very stable and grand "cultural images". Already the word "sustainable development" expresses the tributary to the model of positive change so deeply interconnected to the Western cultural heritage. It affirms the arrow of time. It affirms a hope within this world. It implies that what is unjust can be put right, if only a little more time can be allowed to pass. It gives direction and aim to change. Indeed it is anthropocentric.

This is not the place to go into the possible cultural alternatives. In the history of mankind so many civilizations have varied the different components of such alternatives over and over again.

Here there is only time and place to indicate that other cultures in our times as well as in earlier times have provided drastically different outlooks and thereby operational ways of relating. Other cultures than the Western scientific one have provided examples of being less materialistic and "secular". Some argue that there is a need for Westernized civilization, which has so strongly penetrated almost all corners of the world, to seriously address this issue, if environmental problems are to be reduced. The re-sanctification of the world is one way of giving word to such a thought, which might take many different forms (Nasr, 1968; Hjort & Svedin, 1985).

Maybe such a move would also make it more easy to be less anthropocentric in outlook, reducing our special kind of speciesism. The rights of other creatures have just recently come over the surface and in the economy/ecology interface been addressed, e.g. in the form of biodiversity conservation, basically for the future instrumental value (Aniansson & Svedin, 1990). However, the existence values, which are starting to be discussed, introduce the proper and challenging why, whose, and how much, questions.

The "Western" way of making analytically sharp distinctions between categories, not least the demand that categories should be mutually totally exclusive, has often been attributed to the logic of Aristotle. The positive attitude towards the logic of Aristotle was clearly expressed in the West especially during Medieval times but has strongly influenced the way of approaching an analysis of Nature prevailing in "the West" until our days. Thus we can see a sort of cultural communality in the practice of thought models, in which Aristotelian logic serves as a cultural "Western" code.

The capacity to make sharp analytical distinctions is also the basis for any form of taxonomy, i.e. classification schemes of phenomena in Nature. A classical example is the systematic taxonomy by Linnaeus for plants. This is a major achievement of systems building based on the hierarchically arranged separation of different categories.

Sharp taxonomic distinctions can also be found in Islam. Here very subtle differences between phenomena are given in the Koran text. Man is also different from other phenomena in Nature, especially through the special tasks given to Man by God.

The basic problem, however, is not the classification of already separated categories, but how to choose criteria for a separation of objects which within another approach should not be separated.

Approaches which do not apply such sharp and exclusive distinctions are common in the traditions of the Far East. The question as to whether Man is qualitatively different from or is close to other animals is in that perspective not as problematic as it has been in the West.

The Japanese scholar Kinhide Mushakoji (Mushakoji, 1976) remarked that some aspects of Japanese culture could be understood in terms of a distinction between "*erabi*" culture (from the word for "select" or "choose") and "*awase*" culture (from the word for "combine", but also "to adjust one thing to another"), the latter being more characteristic of the Japanese attitude. The selection-oriented culture emphasizes taxonomic distinctions, whereas the more open "combination" approach never tries to pinpoint a distinction before it has been found necessary.

According to the *erabi* view, ideally Man can freely manipulate his environment for his own purposes, Mushakoji says. This view implies a behavioral sequence, whereby a person sets his objective, develops a plan designed to reach that objective, and then acts to change the environment according to that plan. The erabi view also implies a logical structure composed of concepts and their opposites: hot versus cold, large versus small and so on. When forming a plan of action to change the environment it is, therefore, natural for Man to make a decision in terms of for example 'is it better hot or cold?' Thus, erabi logic is choosing the best option from a set of alternatives composed of selected factors from various dichotomies.

Awase, on the other hand, rejects the idea that Man can manipulate the environment and assumes instead that he adjusts himself to it. According to this way of thinking, the environment is not characterized by dichotomous concepts such as hot versus cold. Rather, it may be somewhat hot or somewhat cold, rather large or rather small. In other words, the environment consists of a constantly changing continuum of fine gradations. Awase is the logic of seeking to apprehend and adapt to these fine gradations of change.

The provocative notion of Gaia (e.g. Lovelock, 1979, 1988), outlining a mechanism of development of our planetary part in the universe addresses, at least implicitly, the question of intent and aim, so strongly suppressed in our western scientific tradition. These issues might reoccur in surprising ways, maybe through the advancement of systemic theory.

In all these cases a careful observer watching the interface region between ecology and economy can already sense the ripples of the changes feared, the changes possible, and the changes necessary in our relations with the natural world.

6. THE CONTEXT OF THE LONG-TERM SOCIETAL SITUATION

So far we have basically dealt with the varying contexts of importance in the *analysis of the environmental problems*. But there is also a *context of reality itself* not connected narrowly to how we perceive or do not perceive a situation. This "context of the long-term societal situation" acts as a forceful provider of the basic "real world" framework, for example the investments already made in technical and social infrastructure and the cultural patterns which direct the momentum of social action into the future.

Thus whatever happens cannot be divorced from this wider societal context. As the impacts of societal action on the environment are connected to a very high degree to this general developmental framework we must include some discussion on this final type of context.

In order to discuss its influence, that is the degree to which sequences of development could be seen as more or less deterministically set by this type of context, we will enter the argumentation from the point of angle of technological change. An interesting starting point is the process of substitution, the process in which, for example, one technological solution takes over from an earlier one. A historical example is when Man changed from wood to coal for heating, and then later on to oil and hydro-electric power, nuclear power, solar power etc. (Figure 1).

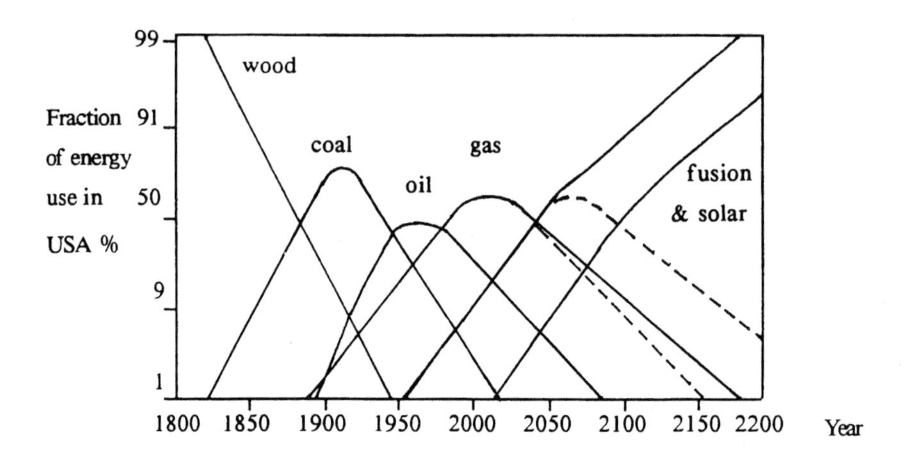

Figure 1. The chain of energy substitutions. The percentage share of different energy sources in US energy use 1800-1975, and projected developments (after Marchetti, 1975).

A substitution between technologies can thus occur in a chain of events where old ones are bypassed by new ones in a distinct succession.

An interesting way of finding out something about the future of such processes used in technology assessment exercises (Martino, 1972) is to extrapolate not the trend of a single technology but the performance of the envelope of successive technologies (Figure 2).

The interesting point for us is that through the use of an envelope-projection we have become a little less sensitive to particular changes in performance (in this case in a specific technology) through the method of searching for a megatrend. This approach normally gives an improved capacity to say something about the future without knowing about the details. This has been gained, however, through the introduction of a subjective element in terms of the choice of selected items which are entered into the analysis (Schwarz et al., 1982). The choice of the proper variable with which to describe a specific technology performance, and which essentially tells you something crucial about future performances, is here of great importance. Is, for example, the number of numerical operations per second for a computer an interesting measure for the future? Or is it the capacity of operations per unit volume? Or is it some sort of measure of the type of operations it can perform? Or is it some measure of self-coding capacity etc? In any case, the choice of an envelope type of projection is in a way, if well done, more interesting than a "naive" single specific technology projection. Why is that so? It is because it integrates over a wider contextual domain.

So, let us take a closer look at these contexts and their possible roles. Let us start with a simple substitution process between an "old" technology (or some old material) and a new one. Substitution of wool as a material for clothing by cotton between 1820

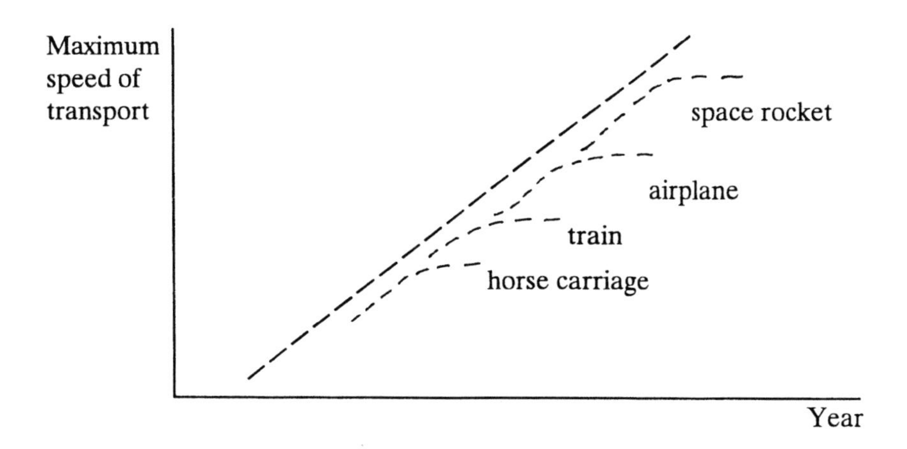

Figure 2. Conceptual presentation of envelop projections of new unknown technological capacities based on historical performances.

and 1870 in Sweden makes the point very clear. Cotton, which was imported from the United States via England to Sweden, grew in importance due to a higher degree of mechanization. Thus, the price for the new material became more competitive as compared with the price for the old material wool. Accordingly, over a period of 50 years there was an almost complete take over by cotton from wool. Historically, the replacement process followed a "logistic" S-shaped curve starting slowly, then accelerating in the middle, and finally settling at the saturation top level.

Such processes have a surprisingly similar pattern. Fisher and Pry (1971) have studied several such historical developments. After proper normalization procedures the "take over" processes fall along the same type of logistic curve[1] (Figure 3).

This is a special case of the so-called Pearl's law, which has been used, for example, in biology (Pearl, 1925).

This was, however, not an entirely new and fresh observation. What was new was that so many substitution processes followed the same pattern. This study included different substitutions as different as that of synthetic rubber versus natural rubber and that of titanium oxide replacing lead-zinc oxide in colour pigments.

The interesting observation is thus that there exists some sort of determinism. Whenever a process of substitution from A to B has started it runs (in these chosen examples) as if it was on rails to an almost complete take-over. However, the data has to be interpreted carefully. In order to get these curves to conform nicely to total overlapping, the authors had to define the "total possible market" in a somewhat subjective manner since the two materials involved in the substitution are rarely so similar that they entirely overlap in performance or quality (Svedin, 1986).

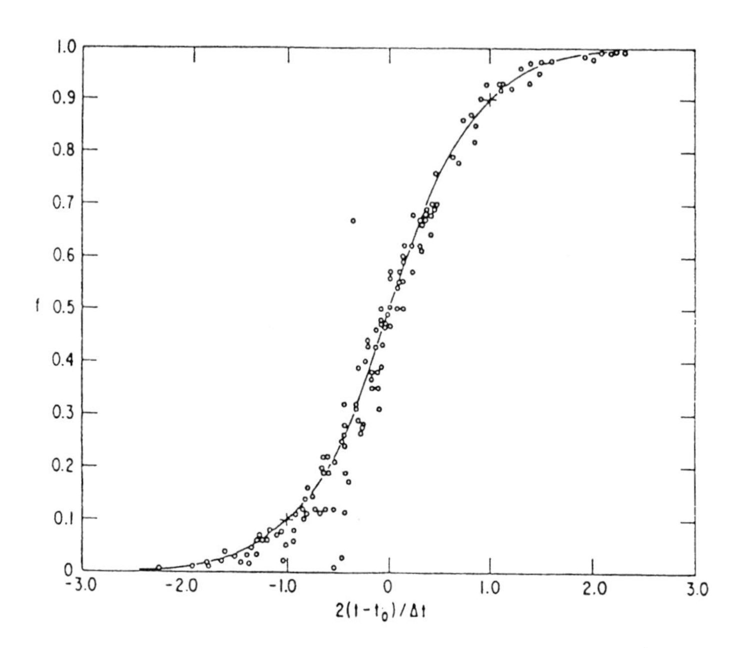

Figure 3. Fit to the Fisher-Pry model of substitution, in terms of a logistic curve, for 17 cases of substitution (all relevantly normalized, e.g. with regard to "starting time" and "tempo of process") (from Fisher-Pry, 1971).

One especially interesting case concerned the substitution of natural by synthetic rubber. If you convert the S-shaped plot to a linear representation (by means of just an ordinary mathematical transformation of the representation) the conformity by which a certain substitution process follows the S-shape is reflected in the extent to which the experimental points follow the straight lines.

Specifically agriculturally based products and their substitution performances are represented in Figure 4. Margarine versus butter and synthetic fibres versus natural fibres both show a remarkable sameness of dynamic change over time (see the two curves on the lower right in Figure 4).

However, it is more interesting to reflect on the take-over curve of synthetic rubber from natural rubber. It starts so oddly, rises sharply and then bends over and after some time smoothly adopts the regular procedure the other curves present following the straight line upwards (corresponding to an adoption at this stage to the S-shape take-over time development). What has happened? A simple scrutiny of the time of appearance of the odd feature shows that it is the 2nd World War event which is reflected. That is of course not surprising at all. The war really forced a material substitution of something urgently needed and available (synthetic rubber) for something desperately lacking (natural rubber). The interesting point is that the curve *after* the war adopts the shape of the other curves *as if nothing upsettning had ever happened*.

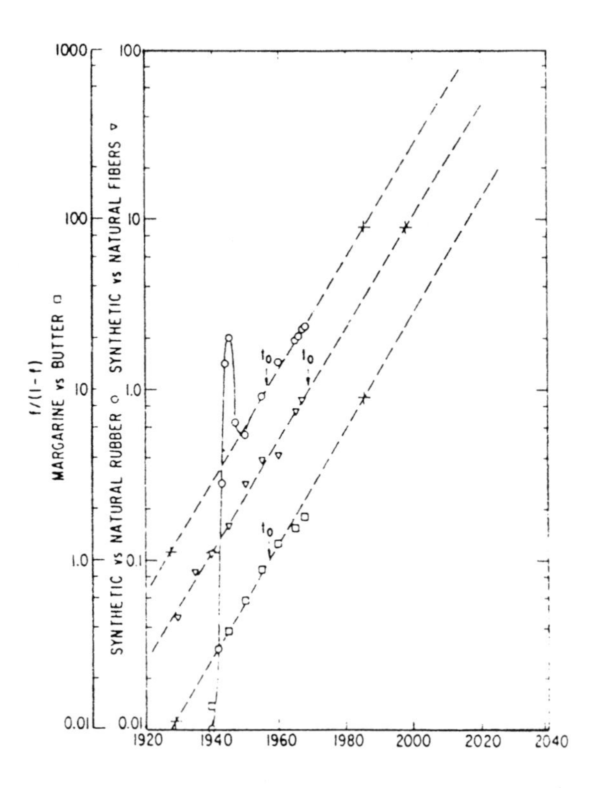

Figure 4. Linear representations of logistic curves for take-over processes according to Fischer-Pry. The full drawn curve corresponds to the take-over of synthetic rubber from natural rubber (from Fisher-Pry, 1971).

The interpretation of this could be made like this: When the war ended, normal market and technological societal contexts took over and these had an inherent stability over time such that in this perspective, the war can be seen just as a minor intermission. In a wider structural context the system exposes a surprising stability over the period 1920 to 1970. Important elements in this integrated world context would speculatively be the existence of a world market, a certain type of capitalistic form of economy on the world scale, a certain type of technological change pattern in terms of relations between fundamental discoveries and the industrial penetration of these.

The point here is not to say definitely that this *is* the full truth for the rubber example enfolded in this larger context and even less to make very bold generalizations. The point is to *exemplify* a possible type of *metastability of conditions* over a certain time span.

Let us now consider another fundamentally interesting case. According to the Fischer-Pry take-over formula, the "new" has deterministically and forever gained over the "old" whenever the process has taken a firm start. The contextual driving forces then have such a power that nothing can prevent the take over from occurring in an almost mechanical manner. Such examples have been shown convincingly, for example for the take over in the energy field between wood and mineral oil for heating purposes, as demonstrated in Figure 1 above.

It is now interesting to contemplate the case where the entirely new technology of biomass production, in a new world of strongly shifted energy price relations and an entirely new ideological political situation (e.g. within the framework of a political decision to abstain from the nuclear option) would present "biomass" as *the new* solution. We would then have a "reversal" of the substitution process, which could be described in two steps (Figure 5).

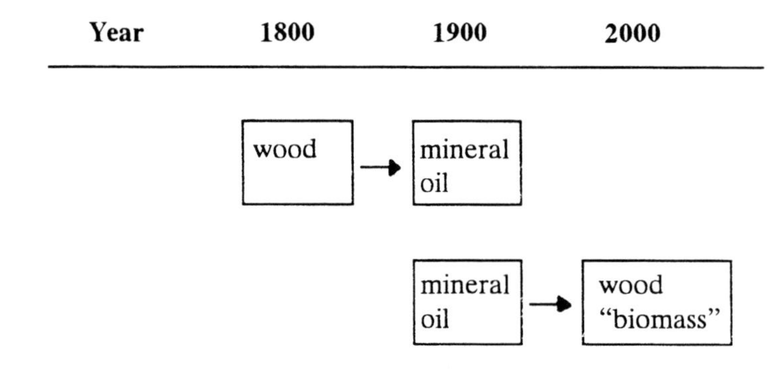

Figure 5. The possible sequence wood-mineral oil-wood (biomass) for fuel use.

According to the "laws" we earlier discussed, the take over would again happen in a mechanical sequence and with a definite touch of determinism.

The interesting point would then, however, be that we have made a loop and in some sense we "are back" to wood. The determinism in the first part of the chain would then seem to be completely disproved. Any segmented path of occurring or recurring technologies could then form any type of chain of events.

The didactic point here is to demonstrate two aspects, the first of which is a remarkable degree of determinism *up to a certain point*. There seems to exist some sort of contextual determinism, but local in space and time. And at the same time there seems to exist a much more open ended situation in the longer term. What made the situation turn back *from* oil and return to wood once more? The interpretation must be that the societal context within which these specific changes take place had also changed to such a degree that the overall situation could no longer be regarded as being the same. And that is very reasonable. The 20th century societal and technological overall context is so different from that of the mid 19th century.

What we have done is thus to transfer our substitution problematique from a question of change of isolated single technologies to changes of vast systemic societal contexts, within which these more specific changes happen as ripples on the sea surface.

What has all this to do with the environment and with the analysis of its problems as seen from the environment-economy interface? First of all, the processes of technological change and its societal conditions have been central topics in economics for a long time. The subtleties and the flexibilities of these processes - with or without semi-deterministic traits - have been carefully studied by economists. Here it is not advocated that these processes of technological change are always deterministic. The point of the argument is didactical in character, i.e. to show that these processes cannot be seen as isolated sequences without any connection with a broader societal context within which they are embedded and which *conditions* their behaviour.

In many ways environmental problems, and also possible future solutions, are technology driven. The understanding that the dynamics of technological change in many ways takes place within a very broad and fairly long term societal context is thus of considerable importance for future challenges of enquiry in the economy-ecology interface. As an example, the impact of biotechnology will be significant in the future. It will most probably create new environmental settings providing both new dangers as well as new options to solutions of old problems. It is thus important to understand the dynamics of the expansion of this technology as part of the future conditions to be studied by both the traditional domains of economics as well as in ecology. And this is only one of several possible examples of this kind. The societal context already exists and it exerts its influence on all these processes.

7. CONCLUSION

In the future the challenge for analysts in the economy-ecology interface will thus not only be to deliberately and carefully disentangle the different perceptual and analytical contexts we started with. This task is not merely, in a reductionist way, to reduce the variations of contexts, but to understand them and link them appropriately in the analysis, depending on what the problem is.

An even more difficult task is to disentangle the layers of long term but constantly changing societal contexts "out there" in terms of a factually existing societal momentum, whatever the reason for this may be: physical infrastructure, financial investments, institutional frameworks or cultural habits. The really intriguing point is that this "outer factually existing context" has to be addressed analytically exactly through the penetration of the type of intellectual contexts we started with.

NOTES

1. If f is the fraction "the new" has taken of the market at a time t, with b as constant, the form of the take-over procedure could be given a mathematical form

   $f = 1/(1+exp(b(t-t_0)))$,

   where t_0 is the time at which the substitution process is just in the middle

   i.e. where $f=1/(2)$

   The constant b can be seen as expressing the initial speed by which the "new" starts to penetrate.

REFERENCES

Allwood, J., Frängsmyr, T., Hjort, A. and Svedin, U. 1981. Natural Resources in a Cultural Perspective. The Council for Planning and Coordination of Research (FRN), Report No. 37S, Stockholm.

Aniansson, B. and Svedin, U. (eds.). 1990. Towards an Ecologically Sustainable Economy. Report from a Policy Seminar, January 3-4, 1990, Stockholm, on behalf of the Environmental Advisory Council of the Swedish Government. Swedish Council for Planning and Coordination of Research (FRN), Stockholm. Report 90:6.

Ayres, R. 1978. Resources, Environment, and Economics: Applications of the Materials/Energy Balance Principle. Wiley-Interscience, New York. Cited in Kates, 1988.

Bertelman, T., Hollander, E., Olsson, C.A., Parmsund, M., Sohlman, S. and Svedin, U. 1980. Resources, Society and the Future: A Report Prepared for the Swedish Secretariat for Futures Studies. Pergamon Press, Oxford. The Swedish original "Resurser, Samhälle och Framtiden, Liber Förlag, Stockholm was published 1977.

Boulding, K.E. 1981. Ecodynamics: A New Theory of Societal Evaluation. Sage Publications, Beverly Hills, California. Cited in Kates, 1988.

Brinck, P., Nilsson, L.M. and Svedin, U. 1988. Ecosystem Redevelopment. *Ambio* 17:88-89

Butzer, K.W. 1982. Archaeology as Human Ecology: Method and Theory for a Contextual Approach. Cambridge University Press, Cambridge. Cited in Kates, 1988.

Clark, W. and Munn, T. (eds.). 1986. Sustainable Development of the Biosphere. Cambridge University Press, Cambridge.

Duncan, O.D. 1964. Social Organization and the Ecosystem. In: Furis, R.E.L. (ed.). Handbook of Modern Sociology. Rand McNally, Chicago. Cited in Kates, 1988.

Ellen, R. 1982. Environment, Subsistence and System: The Ecology of Small-Scale Social Formations. Cambridge University Press, Cambridge. Cited in Kates, 1988.

Firey, W. 1960. Man, Mind and Land. The Free Press, Glencol Ill. Cited in Kates, 1988.

Fischer, J.C. and Pry, R.M. 1971. A Simple Substitution Model of Technological Change. *Technological Forecasting and Social Change* 3:75-88.

Forde, C.D. 1963. Habitat, Economy and Society. E.P. Dutton, New York. Cited in Kates, 1988.

Harris, M. 1979. Cultural Materialism: The Struggle for a Science of Culture. Random House, New York. Cited in Kates, 1988.

Hjort, A. and Svedin, U. (eds.). 1985. Jord, Människa, Himmel: Några Länders Syn på Naturen i De Stora Kulturerna. Liber Förlag, Stockholm (in Swedish). A condensed version exists in english "Earth, Man, Heaven: Philosophical and Religious Views on Nature", (in manuscript).

Holling, C.S. 1973. Resilience and Stability of Ecological Systems. *Annual Review of Ecology and Systematics* 4:1-23.

Holling, C.S. (ed.). 1978. Adaptive Environmental Assessment and Management. John Wiley, Chichester, England. Cited in Kates. 1988.

Holling, C.S. 1990. Integrating Scinece for Sustainable Development. In: Mykletun, J. (ed.). Sustainable Development, Science and Policy. The Conference Report, Bergen. Norwegian Research Council for Science and the Humanities (NAVF), Oslo. pp. 359-370.

Kates, R.W. 1988. Theories of Nature, Society and Technology. In: Baark, E. and Svedin, U. (eds.). Man, Nature and Technology: Essays on the Role of Ideological Perceptions. Macmillan Press, London. pp. 7-36.

LePlay, F. 1879. La Méthode Sociale: Abrigé des Ouvriers Européens. A. Mame, Tours, France. Cited in Kates, 1988.

Leontief, W. 1977. The Future of the World Economy. Oxford University Press, New York. Cited in Kates, 1988.

Lovelock, J. 1979. Gaia: A New Look at Life on Earth. Oxford University Press, Oxford.

Lovelock, J. 1988. The Ages of Gaia. Oxford University Press, Oxford.

Marchetti, C., cited in Rydberg, J. 1975. Kan världen orka med vår levnadsstandard. *Forskning och Framsteg* 8:43-44.

Martino, J. 1972. Technical Forecasting for Decision Making. American Elsevier, New York.

Meadows, D.H., Richardson, J. and Bruckman, G. 1982. Groping in the Dark: The First Decade of Global Modelling. John Wiley, Chichester, England.

Meadows, D.H., Meadows, D.L., Randers, J. and Behrens, W.W. 1972. The Limits to Growth. Universe, New York. Cited in Kates. 1988.

Miller, J.G. 1978. Living Systems. McGraw-Hill, New York

Mushakoji, K. 1976. The Cultural Premises of Japanese Diplomacy. In: Japan Center for International Exchange (eds.). The Silent Power, Japan's Identity and World Role. The Simal Press, Tokyo. pp. 35-49.

Mykletun, J. (ed.). 1990. Sustainable Development, Science and Policy. The Conference Report, Bergen. Norwegian Research Council for Science and the Humanities (NAVF), Oslo.

Nasr, S.H. 1968. Man and Nature: The Spiritual Crises of Modern Man. Unwin Paperbacks, London.

Odum, H.T. 1971. Environment, Power and Society. John Wiley and Sons, New York.

Odum, H.T. 1983. Systems Ecology: An Introduction. John Wiley and Sons, New York. Cited in Kates, 1988.

Passset, R. 1979. L'economique et le vivant. Payot, Paris.

Pearce, D., Markandya, A. and Barbier, E.B. 1989. Blueprint for a Green Economy. Earthscan, London.

Pearl, R. 1925. The Biology of Population Growth. Alfred A. Knopf Inc, New York.

Sachs, I., Bergeret, A., Schiray, M., Sigal, S., Thèry, D. and Vinaver, K. (eds.). 1981. Initiation a l'ecodevelopment. Regard, Privat, Toulouse.

Schwarz, B., Svedin U. and Wittrock, B. 1982. Methods in Futures Studies: Problems and Applications. Westview Press, Boulder, CO 1982.

Svedin, U. 1985. Economic and Ecological Theory: Differences and Similarities. In: Hall, D.O., Myers, N. and Margaris, N.S. (eds.). Economics of Ecosystems Management. Dr. W. Junk Publishers, Dordrecht. pp. 31-39.

Svedin, U. 1986. Comparative Method, Trend Extrapolation, Future Surprises and Societal Context. In: Sällström, P. (ed.). Parts and Wholes: An Inventory of Present Thinking About Parts and Wholes. Vol. 4 Comparative Methods. Swedish Council for Planning and Coordination of Research (FRN), Stockholm.

Svedin, U. 1988. The Concept of Sustainability. In: The Stockholm Group for Studies in Natural Resources Management. (eds.). Perspectives of Sustainable Development: Some Critical Issues Related to the Brundtland Report. Stockholm Studies in Natural Resources Management No. 1, Department of Systems Ecology, Division of Natural Resources Management, Stockholm University, Stockholm. pp. 5-18.

Svedin, U. and Aniansson, B. 1987. Surprising Futures: Notes from an International Workshop on Long-Term World Development. Swedish Council for Planning and Coordination of Research (FRN), Stockholm. Report 87:1.

Svedin, U. and Heurling, B. 1988. (eds.). Swedish Perspectives on Human Response to Global Change. Swedish Council for Planning and Coordination of Research (FRN), Stockholm. Report 88:3. see especially Landberg, H. & Svedin, U. The Societal Dimension of the Environmental Issues: Growing Concern. pp. 7-12.

WCED, World Commission for Environment and Development. 1987. Our Common Future, Oxford University Press, Oxford.

Linking the Natural Environment and the Economy; Essays from the Eco-Eco Group,
Carl Folke and Tomas Kåberger (editors)
Second Edition.
1992. Kluwer Academic Publishers

CHAPTER 2

Actors, Roles and Networks: An Institutional Perspective to Environmental Problems

by

Peter Söderbaum

Department of Economics
Swedish University of Agricultural Sciences
Box 7013, S-750 07 Uppsala, Sweden

To begin with the nature of environmental problems is discussed, for instance the irreversibility of many impacts, the uniqueness of natural resources, and uncertainty. The institutional aspects of environmental problems are also indicated. A conceptual framework, an actor's perspective, with concepts such as actor, role, network, power and resources, institutions, values and ethics, is then suggested and compared with recent public choice theories, the latter forming the part of the conventional economic paradigm. It is argued that neo-classical economics largely avoids important aspects of environmental problems, for instance the ethical issue, and that economic research based upon unconventional perspectives is more promising.

1. INTRODUCTION

Humanity is facing a number of problems relating to development and the environment. Some problems are of a local character, others are national or even global in nature. Perception of such problems is largely a matter of the conceptual spectacles that are used. All actors on the public scene whether politicians, bureaucrats, journalists, environmentalists or business leaders use some kind of conceptual framework that directs attention in specific ways.

Social scientists, and economists more than others, claim to know something about useful conceptual frameworks in human problem solving. And economics in western countries today is largely synonymous with the dominant neoclassical school. One may well ask, therefore, whether some modified version of conventional economics is all that is needed to cope with the problems we are facing.

It is true that neoclassical economics has something to offer in terms of conceptual frameworks at the micro or macro level. But neoclassical economics is just one perspective and, depending upon the problems faced, some other perspective or perspectives may well be more fruitful. In some cases, and I suggest environmental problems are a good example, the conventional approach may even be part of the problem. Neoclassical economics could therefore aggravate rather than reduce the problems.

Such a statement is based upon my belief that certain other theories and conceptual frameworks are more useful. And our relationship to alternative schools of thought or conceptual frameworks is always a matter of values and beliefs.

In the following pages I shall emphasize environmental problems. I hope, however, that much of the discussion will also be relevant to other kinds of development issues.

2. THE NATURE OF ENVIRONMENTAL PROBLEMS

Formulating the problems to be dealt with is a step that necessarily involves some manipulation. One way to proceed is to take the neoclassical framework for granted and describe the problems accordingly in terms of externalities, marginal costs and benefits of pollution abatement etc. A more many-sided approach reflecting the holistic ambitions of *institutional economics* suggests the following characteristics of environmental problems as a suitable point of departure.

- The problems are multidimensional and multidisciplinary rather than one-dimensional or limited to one discipline.
- The problems are non-monetary as well as monetary in nature.
- Any degradation of non-monetary resources such as human beings, social relationships, ecosystems, natural resources is often irreversible or very difficult to reverse.
- Non-monetary resources are often unique or very rare either in an objective or in a subjective sense.
- The problems often extend beyond the actors involved in market transactions,

i.e. impinge on interests such as the property rights of other parties.

- The problems often have considerable spatial extension, e.g. they are not limited to one single municipality, county or country.
- The problems are multisectoral, i.e. not limited to one sector of society or the economy.
- The problems involve uncertainty and risk (where only part of the uncertainty can be reduced through research efforts or the acquisition of knowledge available elsewhere in the economy).
- The problems involve conflicts between different interests and ideologies in society.

This list of characteristics has been discussed in more detail in connection with attempts to design approaches to decision making other than the conventional cost-benefit analysis (Söderbaum, 1986, 1987). The list in itself indicates that there are problems related to our cognitive habits. Trying to deal with non-monetary problems, irreversibility, uniqueness etc. in monetary terms (by using cost-benefit analysis) may not be the best strategy. For our present purposes, it is important to observe the complex character of environmental problems. Many aspects are involved. Our knowledge is fragmentary rather than complete and uncertainty about technical options and impacts is the normal case rather than the exception.

The current focus is, however, on actors and the institutional framework and a broader indication will therefore be given of which problem areas are involved.

Low performance in one region or country with respect to environmental indicators - such as the state of the soil, forests, lakes or groundwater - could perhaps be related to each of the following factor categories, or rather to a specific combination of these interrelated factors:

- Cognitive habits, language, theoretical frameworks.
- Knowledge and information.
- Values, ideologies, ethics.
- Organization, cooperation and conflicts.
- Power and control of resources.
- Behaviour styles, professional styles, roles and strategies of different actor categories.
- Institutional context and incentive systems.

I have already referred to the first factor in the list, i.e. the possibility that our thinking habits relating to economics and management of resources in business firms or at the national level may be part of the problem. Neoclassical economic theory (public choice theory included - see below) may legitimize thinking habits that are detrimental to the environment and thereby to the natural resource base available for future generations. Notwithstanding the debate that has been going on for some time, GDP-growth still seems to be the main idea of national progress held by many influential actors such as politicians and business leaders. And the monetary conception of resource allocation (which I regard as monetary reductionism) still predominates at the micro level.

In many cases our knowledge and information is fragmentary, contradictory and insufficient. With thousands of chemicals around, with all their possible synergetic effects, one may well ask whether it is possible at all to come to grips with the situation. Monitoring and decision support systems used for the control and guidance of business firms or national economies are generally designed for the purpose of monetary rather than environmental performance. Monetary survival and prosperity, with some side reference to employment, seems to be the overall consideration, whereas survival in a biological or ecological sense is downgraded or simply disregarded.

Combining values and factors related to knowledge in the list given, one may well discuss the environmental awareness not only of specific actors but also of the public at large. Many people who are attracted by the idea of the welfare state believe that environmental problems, like all other problems, are well handled by governmental agencies and bureaucrats. They believe that civil servants look after the public interest. I will return to this question of professional roles in my discussion about public choice theory and its alternatives.

Similarly, many of the institutions, i.e. rules of the game and incentive systems that have evolved or been designed over the years, do not sufficiently reflect environmental values. In many cases current rules benefit Big Business and Big Government at the expense of smaller property holders, citizens or environmental organizations. Often these rules protect traditional power structures more than they protect the environment, if that comparison may be permitted. In any society, traditional power structures will only be changed gradually and in the present case such changes will depend upon the importance that citizens attach to conservation and environmental values. But the processes of social change are rather complex, as we will see. Where a majority of politicians agree to change laws, bureaucrats may well be slow in implementing the measures or even interpret things in their own way.

3. INSTITUTIONAL ECONOMICS

What then are the alternatives to conventional economics of the neoclassical kind? Broadly speaking, there are three different schools of thought, Marxian economics, institutional economics and neoclassical economics (see for instance Fusfeld, 1986). In addition there are some other non-conventional groups such as those who speak of social economics or humanistic economics. Also there are subdivisions of "mainly neoclassical" economists such as Keynesians and members of the Austrian School. Advocates of public choice theory could also be regarded as such a subdivision within the neoclassical tradition.

All three schools that I have referred to have some of their intellectual roots in the classical economics of Adam Smith, David Ricardo and others. Marxian economics is sometimes referred to as radical economics in the USA, the main journal being the *Review of Radical Political Economics*. My preference, however, is institutional economics, which relates back to American institutionalism of the late nineteenth and early twentieth century and before that, in some respects, to the German Historical school. Thorstein Veblen, John R. Commons and Wesley C. Mitchell are generally regarded as the founders of American institutionalism. Today there are many followers organized in the Association for Evolutionary Economics (AFEE), with its *Journal of Economic Issues (JEI)*.

Contemporary American institutionalists include, for example, Allan G. Gruchy, Philip A. Klein and Marc R. Tool, editor of the *JEI*, and as with other schools of thought some evolution in conceptual framework and emphasis has taken place, among other things to reflect the public policy issues of more recent times.

A number of European scholars have also regarded themselves as institutionalists. In connection with environmental issues, I would like to mention K. William Kapp from Switzerland (Kapp, 1950, 1970, 1976). Another person who has influenced me considerably is Gunnar Myrdal, who started out as a neoclassical economist (being a member of the "Stockholm School" of economics), but who later gradually changed his perspective in an institutional direction (Myrdal, 1972, 1978). Referring back to the work of Kapp and Myrdal among others, a European Association for Evolutionary, Political Economy was formed in June 1988.

Allan Gruchy, and also Myrdal, have emphasized the holistic character of institutional economics (as opposed to the atomism or reductionism of large areas of neoclassical economics). An interdisciplinary orientation is another characteristic of institutionalism, or perhaps just an example of the holistic ambition.

Gunnar Myrdal is probably best known for his criticism of the alleged value-neutrality and positivistic character of neoclassical economics. According to Myrdal, *Values are always with us ... There is no view except from a viewpoint.* In the questions raised and the viewpoint chosen, valuations aıe implied (1978 pp. 778-779).

Myrdal and other institutionalists therefore share the recommendation of Marxian economists that economics should be called "political economics" to point to the inherent political character of all economic theories. This does not mean that economists should become politicians but rather that we should be conscious of the valuational aspects of our work, explicitly state the valuational premises of our research and perhaps - this is my own suggestion - wherever possible *consider several valuational viewpoints rather than just one*.

It should be noted that Marxian and institutional economists are not the first to speak of political economics. This was the vocabulary for a long time starting with the classical economists and, as is well known, Adam Smith also held a chair in moral philosophy. When "political economics" was changed into "economics" or "pure economics" in the latter part of the nineteenth century, this could well be seen as a step backwards. An increasing number of economists have now come to realize that the attempt to construct a value-free economics in Newtonian equilibrium terms and a variety of simplistic mathematical models was a partial failure. Much of this so-called sophisticated economics that still dominates in mainstream journals is simply irrelevant. Other parts of it may bear some relevance, but are so specific in their assumptions that they cannot be presented as the basic principles of our discipline.

4. A CRITIQUE OF PUBLIC CHOICE THEORY

Some instances of simplistic or reductionist tendencies in mainstream theory relate to human motives. In traditional microeconomic theory, it is assumed that firms maximize profits and nothing more. Improvement or deterioration of non-monetary resources is not considered in a conceptually satisfactory manner and ethical issues about the "social responsibility of business" or the need for "social accounting" are simply disregarded.

Similarly, more recent public choice theories extend the notion of "rent-seeking" behaviour to all kinds of interest groups and actors on the public scene. Just as farmers maximize their profits and other incomes, bureaucrats maximize the budget that they control, thereby increasing their prestige and improving their power position (Niskanen, 1971), and politicians seek to maximize the number of votes. Such theories have been taken seriously among mainstream economists in recent years and one of the advocates of such a theory, James Buchanan, has even received the Nobel prize for economics.

It is true that these theories in some respects represent a step forward for mainstream economists. While institutional economists have pointed to the importance of institutional arrangements for more than one hundred years, one can only welcome a situation in which this interest is also shared by mainstream economists.

The assumption that narrow egoistic motives influence the behaviour of different actors such as politicians and bureaucrats also seems realistic. In this way, naive ideas of bureaucrats or politicians as actors with *only* altruistic motives are challenged. But the idea that man is *exclusively* guided by egoistic motives seems equally dubious. It seems more realistic and fruitful (in a normative manner) to assume that man is ruled by both egoistic *and* more altruistic motives and that the "proportions" of these components vary with time for one person and vary between individuals. This means that there is also cause to question the assumption about homogeneity of motives within each actor category (farmers, bureaucrats, politicians etc).

Consider the sentence: "Farmers are motivated solely by egoism and they are all alike." Within a positivist tradition such statements are regarded as neutral with respect to values, the idea being that they can be tested with respect to truth. But all theories have a normative content and their function or role in society is not restricted to explanation or prediction. A specific theory such as the public choice theory or "the" theory of the firm may legitimize or justify certain types of behaviour among bureaucrats or business executives. A theory that assumes that certain actors are narrow-minded will not do much to broaden the motives and thinking habits of these actors.

In this respect, public choice theory in addition to being questionable from an explanatory point of view is also dangerous to society. The theory not only explains but also influences behaviour to make it less incompatible with the theory. This means that the theory is used as a political instrument.

Theories based on the idea of rent-seeking behaviour rule out issues of ethics as unimportant or irrelevant. Questions about intra- and intergenerational justice or other aspects of ecological ethics become meaningless within the scope of such a theory. Those who believe that narrow motives and reductionistic thinking are good for society certainly welcome public choice theories. Actors and citizens who consider a broader spectrum of human motives and favour holistic thinking will look for other theories.

These other theories certainly also have some political content. We may then conclude that predilection for one specific theory is not only a matter of scientific judgement.

5. ACTORS, ROLES AND NETWORKS: A SOCIAL CHANGE PERSPECTIVE

As I have already indicated, institutional economists have an interdisciplinary orientation and are open-minded about the possibility of learning from other social sciences such as social anthropology, sociology, psychology, business administration (e.g. organization theory), political science etc. Concepts such as perception (i.e. selective perception), cognition, attitude, role, social systems and social networks are judged also to be relevant to us as economists.

For our present purposes I wish particularly to emphasize the following concepts:

- Actor
- Role (each actor having ideas about his own roles and those of other actors)
- Strategy (conception of means and ends from the point of view of specific actors or organizations)
- Organization, social network (the latter bringing together actors from one or from different actor categories)
- Institution (referring to rules of the game which may be formal, such as laws, or informal)

- Arena or stage (i.e. places or social contexts where actors meet)
- Agenda (issues brought to public attention or to the attention of specific decision makers)
- Power and resource position (of actors, organizations or networks)
- Values, ethics, responsibility of actors in their different roles.

In a specific society, or rather in relation to a specific problem area, such as the prevention of cancer, many actor categories can be identified. There are politicians, bureaucrats and businessmen as well as journalists, researchers, educators, lawyers, representatives of trade unions and public interest groups.

Each individual plays many roles. In addition to a professional role or the role of wage-earner and consumer, he or she may engage in many kinds of activities, such as recreational activities, involvement in trade unions or public interest organizations (e.g. environmental groups), or political activity (e.g. as a member of a political party).

An individual may experience certain incongruences between his or her different roles. A businessman, for instance, may at the same time be an organized member of the nature conservation movement and, as a professional, responsible for discharges of pollutants into a nearby lake or river. Any feelings of tension or dissonance in such situations may be counteracted by attempts to modify behaviour connected with specific roles. In this way the individual may succeed in improving congruence or move some way towards a clearer identity. This identity or self-image is connected with some idea about "what direction is forward." Each person has a specific orientation with respect to cognition (thinking habits) and values. We may refer to this orientation as a specific "ideological orientation."

There are certainly important similarities between individual members of one actor category (e.g. between farmers *or* between bureaucrats), but there are also differences. When discussing scope for social change, an assumption of heterogeneity within each actor category with respect to motives and behaviour may be regarded as more fruitful than the homogeneity assumption of public choice theories.

Each individual experiences a number of institutions or rules. Only a sub-set of these rules is connected with markets for the exchange of goods and services. Monetary and non-monetary rewards and punishments are connected with the acceptance or non-acceptance of the rules perceived (Figure 1).

Monetary rewards and punishments may be exemplified by increases in income and the absence of such increases. Being accepted as a member of a specific social group or organization exemplifies a non-monetary reward, while non-acceptance in a group is a case of non-monetary punishment. At this level of the individual, psychological learning theory (suggesting, for instance, that behaviour which is rewarded will be reinforced) is relevant to an understanding of behavioural changes. But various ideas about psychological defence mechanisms and their suggested explanations could also be integrated into our model of individual actors.

	Rewards	Punishments
Monetary	I	II
Non-monetary	III	IV

Figure 1. Specific behaviour alternatives are followed by rewards and/or punishments and these may be of a monetary or non-monetary character.

Change can be initiated by individuals acting in different roles. The role of consumer, for instance, is only one of several possible roles. Individuals can cooperate in organizations or in social networks connecting individuals within or across actor categories. Individuals concerned about environmental problems may be found not only in environmental groups but also among politicians, journalists, scientists, educators, farmers etc. In this way a kind of collective action which differs from the one discussed in the public choice literature (cf. Olson, 1965, 1982) becomes important in understanding social change (Söderbaum, 1989).

In a specific national or local society, opposing pressures for change may exist side by side. Actors and organizations with their networks working for change in one direction may compete with other networks opposing change or holding other ideas about societal development. The struggle or competition may concern language or terminology, development concepts or visions (eco-development versus simplistic GDP growth), business concepts, ideas about lifestyles, rules of the game (direction of institutional change), resources and power positions and, of course, other social and physical changes in society.

Change may be initiated, accelerated or hindered by state legislation. But the conceptual scheme suggested also opens up the possibility of decentralized initiatives, decisions and actions. Formal rules of the game in terms of laws, guidelines etc. often represent an *ex post* adaptation to the informal changes that have already occurred.

6. EMPIRICAL STUDIES

As an example of a study that has been carried out broadly along the lines suggested here, I would like to point to Samuel Epstein's book *The Politics of Cancer* (1978). Starting with certain substances which are now judged to be carcinogenic (for instance pesticides such as Aldrin, Dieldrin, Chlordane, Heptachlor), he goes back to the origins of the debate, with cases of damage, scientific reports, initiatives by labour unions, scientists, public interest groups etc.

Epstein's findings from the U.S.A. support my suggestion that there is a heterogeneity among scientists as a group or among business leaders as a group. Epstein's analysis also raises certain issues related to ethics, such as the social responsibility of business or the responsibility of various professional groups. According to Epstein, for instance, establishment groups such as the American Cancer Society have done very little or nothing to increase public awareness about the risks connected with the substances under study. Citizen groups such as EDF (Environmental Defense Fund), HRG (Health Research Group), NRDC (Natural Resource Defense Council) and in some cases (vinyl chloride) labour unions have been more successful.

Other examples of relevant studies concern energy systems in Sweden, where environmentalists have a different idea about priorities in terms of lifestyles and technology from that of power supply companies. Attempts have been made to make each actor category better understand the motives and cognitive orientation of the other party as a first small step towards reconciliation (Sjöström, 1985).

7. THE ROLE OF THE BUREAUCRAT: A DISCUSSION

Let us now return for a moment to our bureaucrats. In addition to the budget-maximizing bureaucrat several possible role interpretations may be considered. Other pure categories include:

- The bureaucrat as a *devoted analyst* or person emphasizing allegiance to specific politicians or a political party
- The bureaucrat as an interpreter and *defender of specific interests* (e.g. concerning the environment), negotiating with advocates of other interests
- The bureaucrat who weighs all kinds of impacts and interests against each other, applying his professional skills, and points out the *optimal solution*
- The bureaucrat who emphasizes *democracy* as a meta-institution and rule of conduct for himself and other actors. Responsibility and responsiveness are key features of this role conception and a many-sided public debate of alternatives and values is facilitated rather than hindered.

The relative frequency of such role conceptions (and of various combinations of the motives indicated) has to be established through empirical research rather than armchair model building. But as I see it, the main thing is to raise the normative question about the most desirable role conception from a societal point of view. What bureaucrats do we need to assist in the handling of various natural resource problems?

There is no doubt that, my own preference is for the last of the above role interpretations. In a similar way, democracy is a helpful concept in the discussion of other actor roles, such as that of university research workers. In their choices of strategies in relation to other groups, all actors have to look for and consider meta-rules that facilitate communication and dialogue.

8. TOWARD A RESEARCH AGENDA FOR ECO-DEVELOPMENT

Many or all of the factor categories that were identified initially as potential explanations of environmental performance would be regarded as irrelevant by neoclassical economists or as being of little interest to the discipline of economics. Mainstream economists do not seriously consider alternative conceptual frameworks or paradigms.

The neoclassical assumption of perfect knowledge and information represents a strong tradition, although in fact it is also challenged by some mainstream economists, such as Friedrich von Hayek and George Stiegler. How does that assumption relate to the thousands of chemicals marketed in various countries throughout the world? Rather than simplifying our models we should perhaps try to bring the situation back under control by radically changing the practical use of these chemicals in some areas.

The neoclassical idea that values and "tastes" can and should be taken as given is very different from our proposal that economists should assist in attempts to articulate different ethical or valuational viewpoints. The "problem" is perhaps not to employ given values but rather the values, visions and development concepts themselves.

The neoclassical idea that man is essentially a consumer and that it is the consumer's values that count is very different from our discussion of many roles for each individual, possible conflicts between different roles etc.

The neoclassical idea that prevailing power structures and institutions can be taken for granted is very different from the idea that power and institutions may represent important aspects of the problems faced. It is true that public choice theory represents an attempt to deal with institutions. Unfortunately this attempt seems to be a partial failure.

In conclusion, neoclassical economics tends to"deal with" large areas of environmental issues by avoiding them. A research agenda for eco-development has to face all aspects of the problem: the paradigm problem, the problem of knowledge and lack of knowledge, the problem of professional roles, lifestyles, business concepts and development concepts for society at large, the problem of power relationships and institutional design, i.e. desirable rules of the game should we wish to come to grips with environmental issues.

Rather than modifying details in the conventional approach we should therefore as economists call for a major reallocation of financial and "man-power" resources from conventional economics to the exploration of non-conventional approaches and paradigms. As with all other proposals to reallocate resources, this one is based upon certain values and values can always be challenged in a democratic society.

ACKNOWLEDGEMENT

I have benefited from comments by Richard B. Norgaard on an earlier draft of this paper.

REFERENCES

Epstein, S.S. 1978. The Politics of Cancer. Sierra Club Books, San Francisco.

Fusfeld, D.R. 1986. The Age of the Economist. Scott, Foresman, Glenview, Ill.

Kapp, K.W. 1971 (1950). The Social Cost of Private Enterprise. Shocken, New York.

Kapp, K.W. 1970. Environmental Disruption: General Issues and Methodological Problems. *Social Science Information* (International Social Science Council) 4:15-32.

Kapp, K.W. 1976. The Nature and Significance of Institutional Economics. *Kyklos* 29:209-231.

Myrdal, G. 1972. Against the Stream. Critical Essays on Economics. Random House, New York.

Myrdal, G. 1978. Institutional Economics. *Journal of Economic Issues* 12:771-783.

Niskanen, W.A. 1971. Bureaucracy and Representative Government. Aldine Press, Chicago.

Olson, M. 1965. The Logic of Collective Action: Public Goods and the Theory of Groups. Harvard University Press, Cambridge, Massachusetts.

Olson, M. 1982. The Rise and Decline of Nations: Economic Growth, Stagflation and Social Rigidities. Yale University Press, New Haven.

Sjöström, U. 1985. Låna varandras glasögon. Om energiproduktionen och människans villkor. Pedagogiska institutionen, Stockholms universitet, Stockholm.

Söderbaum, P. 1986. Economics, Ethics and Environmental Problems. *Journal of Interdisciplinary Economics* 1:139-153.

Söderbaum, P. 1987. Environmental Management: A Non-traditional Approach. *Journal of Economic Issues* 20:139-165.

Söderbaum, P. 1989. Environmental and Agricultural Issues: What is the Alternative to Public Choice Theory? In: Dasgupta, P. (ed.). Issues in Contemporary Economics Vol. 3: Policy and Development. International Economic Association and Macmillan Press, London (forthcoming).

Linking the Natural Environment and the Economy; Essays from the Eco-Eco Group,
Carl Folke and Tomas Kåberger (editors)
Second Edition.
1992. Kluwer Academic Publishers

CHAPTER 3

Economic Analysis of Environmental Impacts

by

Jan Bojö

The World Bank
Africa Technical Department, Environment Division
1818 H. Street N.W., Washington D.C. 204 33, USA

The article discusses how economics can be applied to the analysis of environmental problems, with particular emphasis on cost-benefit analysis. It is argued that economics can play an useful role, provided it is sensibly linked to an understanding of ecological processes and the institutional environment that affects decision makers, as either individual resource users or politicians.

1. INTRODUCTION

When Johan Åshuvud and a group of friends initiated the "Eco-Eco Group" in Stockholm, it was in the spirit of an interdisciplinary openness and curiosity that is too rarely encountered. This paper is written with that spirit in mind.

The ideas discussed in this paper are not presented as “the solution” to environmental problems. They are merely examples of constructive contributions that can be made by economists to facilitate rational decisions about the environment. However, economists alone can achieve little. There must be cooperation with other disciplines in order to supply analysis with good data. Natural scientists will have to provide guidance for the identification of relevant environmental impacts and quantitative estimates of their magnitude. Social scientists can contribute an understanding of attitudes, customs and institutions, all of which influence the behaviour of economic agents. Effective communication with decision-makers is another area where economists need help.

The purpose of this paper is *to provide a brief overview of economic methods that can assist the analysis of environmental costs and benefits*. While an economy-wide outlook on environmental impacts is quite important, this paper focuses on project level analysis using cost-benefit analysis (CBA), and is concerned with the *physical* (versus the social) *environment:* air, climate, land, water and vegetation. From an economic point of view, the physical environment is a source of raw materials and energy, a source of amenities (such as recreation, beauty) and an assimilator of waste from consumption and production processes.

The paper is organized as follows. Section 2 discusses briefly why markets and policies fail to adequately address environmental problems. This points to the need for economic analysis to correct for such failures. Section 3 points out a number of reasons why it is important to consider the macro-level perspective. However, the main thrust of this paper is contained in section 4, where project level analysis is discussed. Particular attention is devoted to economic valuation of environmental impacts. Finally, there is a brief discussion about "alternatives" to CBA, and the criticism against its use.

2. ENVIRONMENTAL DEGRADATION AND THE NEED FOR ECONOMIC ANALYSIS

Much environmental degradation is the result of large numbers of individuals engaging in destructive (but privately rational) actions: The gathering of fuelwood while planting nothing in its place, since the tree would not belong to the planter anyway; the temporary tenant who maximizes crop yield without any concern for the conservation of soil; the car driver who uses his vehicle without concern for emissions and noise pollution, and the plant manager who discharges untreated effluent into water and allows emissions into the air because the damage costs do not appear on his financial accounts. These are examples of perfectly understandable behaviour - given a particular economic environment. The crux of the matter is that private rationality and social rationality do not coincide.

When spelling out the need for economic analysis, the point of departure is the *neoclassical*[1] model of an economy with *complete and perfectly competitive markets.* This is not because this model is a good description of the real world, but because it is an analytically clear bench-mark. Deviations from this model can be grouped into two categories: *market failures* and *policy failures*.

Obviously, the real economy, especially in developing countries, is far removed from the neoclassical model economy. Therefore, it would be naive to suggest that every move towards achieving such a model economy is beneficial, given the prevalence of other distortions. The real world is one of "second-best" solutions. Each policy or project needs to be scrutinized on its own merits, and not simply supported because it moves the economy towards greater use of unregulated markets.

An important category of market failures that contributes to environmental degradation is *externalities*, factors external to the existing structure of markets. The pollution of a river without compensation to fishermen losing their income, is one example. Another is the lowered quality of grazing land, affecting all cattle owners, as a result of private decisions to increase stock. In summary, existing markets are often unable to communicate environmental damage in terms of price signals back to the agent causing the damage.

An additional concern with market solutions is the *distribution of income*. Even in a hypothetical situation with complete and perfect markets, one may not be satisfied that the outcome is equitable. How much the inequity is due to the mechanics of the markets and how much is due to other factors is a controversial subject. Clearly, power, inheritance, skill, intelligence and even violence all play their role in the distribution of assets in a society.

Market failures (as well as other social forces) give rise to government interventions, which in turn often result in *policy failures*. Examples include international trade policy barring free trade, price policy depressing incentives for sound development, bureaucratic sub-optimization emphasizing the interests of the elite, misdirected and wasteful investments, and suppression of information regarding environmental mismanagement.

The environmental damage caused by market and policy failures can be analysed by a combination of economics and other means on three levels:

1. The *general policy level*, where the links to environmental damage may not be that obvious, but nevertheless at times they are quite strong. An important example is the definition of property rights within the economy. Obviously, private motivation for investing in soil conservation, tree planting and grazing management will be affected by the degree to which it is possible to harvest the returns on the investments.
2. The *environmental policy level*, where conscious decisions are made to limit environmental degradation through measures such as complete prohibition of certain commodities or emissions, environmental minimum standards, taxation, fees, deposits, subsidies and perhaps tradable discharge permits.
3. The *project level*, where adjustments can be made to optimize[2] environmental damage. The emissions of polluted air or water from industry can be reduced, roads designed so as to limit erosion, the use of agricultural chemicals limited by substitution of biological pest management, health impacts lowered through proper training of personnel, sewage treatment and recycling built into city planning, and so on.

These three levels of economic analysis of environmental problems should be seen as complementary: one without the other may not do much good. It is impossible to reach all firewood gatherers, farmers, vehicle drivers or plant managers at the project level. Specific projects must therefore be complemented and supported by sound general economic policies.

3. ECONOMIC ANALYSIS AT THE NATIONAL LEVEL

The national[3] level of policy-making concerns both the general and specific environmental policy. This is of relevance for economic analysis of environmental impacts for several reasons.

First, *national and regional policies* (e.g. agricultural or trade policies) that are not directly aimed at environmental issues, will often through the repercussions in the economy have a substantial *impact on the environment*. Consider subsidies to beef ranches, that were meant to decrease the price of meat, but that will instead encourage deforestation and flooding of lowlands. Or a tax on export crops that will discourage environmentally beneficial agroforestry schemes that were based on export crops.

Second, economic analysis of a set of environmental problems can be used to promote optimization, for example minimizing the necessary costs for achieving an environmental improvement. Consider the problem of finding the least cost measures to achieve a 30 per cent reduction in sulphur dioxide emissions. A *general equilibrium model* (covering the entire economy) *can be used to derive optimal solutions*. If interactions between markets are more limited, a partial economic model may be sufficient.

Third, a particular *project* may have important *repercussions for the rest of the economy*. Consider a major chemical plant that will make cheap pesticides available for widespread use. This may raise harvests, bring grain prices down and increase food exports. On the other hand, fisheries may be damaged by pesticide runoff, potable water may be poisoned, and so on.

Finally, there is a danger in viewing each project in isolation. Many *separate projects* that appear worthwhile - other things assumed unchanged - may *add up* to a development that is undesirable. Several separate logging operations may combine to destroy a viable habitat for endangered species, for example. Thus, a broader economic analysis than the narrow project scope is sometimes necessary to reach valid conclusions.

4. ECONOMIC ANALYSIS AT THE PROJECT LEVEL

Environmental impacts at the project-level can be analysed through many approaches. Some are labelled *specification techniques* (Cooper, 1981). These organize information about environmental impacts in a consistent but multi-dimensional framework using e.g. EIA (Environmental Impact Assessment; Clark et al., 1978). This is often a good starting point for economic analysis.

The techniques discussed in this paper can be labelled *evaluation techniques*. They aim to synthesize the different dimensions of an environmental problem into a common unit. For practical reasons, this unit is generally money, although any other widely known and accepted yardstick (shells, camels, cattle ...) would also do.

Given the fact that so much of the economy - even in developing countries - is monetized, it appears practical to use it as a unit of measurement. This should not be misunderstood as giving money per se a particular value, or a value superior to everything else:

> "... what is worse, and destructive of civilization, is the pretence that everything has a price or, in other words, that money is the highest of all values." (Schumacher, 1973, pp. 37-38).

Another unit that has been suggested as the unit of account is energy (Costanza, 1984; Odum, 1984). It remains unclear how policy could be derived from such measurements. Obviously, there is no direct link between human welfare and embodied energy; the fact that commodity X has twice the embodied energy of commodity Y has nothing to do with my preferences for one or the other. Thus, energy has to be valued somehow according to its usefulness - which appears to take us back to prices and money.

4.1 Cost-Benefit Analysis and its Major Steps

The core method for environmental economic evaluation is *cost-benefit analysis* (CBA).[4] This is defined here as a process containing the following components:

1. Identification and quantification of social advantages (benefits) and disadvantages (costs) in terms of a common monetary unit.
2. Benefits and costs are primarily valued on the basis of individuals' *willingness to pay* for goods and services. This goes for both market and non-market items (cars as well as clean air).
3. The flow of benefits minus costs over time are brought together into a single value "today" (net present value).
4. Unquantified effects (intangibles) are described qualitatively and compared with quantified values.
5. Discussion of policy implications.

CBA is a normative exercise aimed at providing policy-relevant conclusions, aiding, but not dictating, environmental decision-making. The idea that results emanating from CBA are somehow "objective" is not supported here. *CBA is based on two assumptions:* that *individual preferences should count*, and that *individuals are generally the best judges of their own welfare.*

Obviously, one is free to argue that individuals should not be allowed to make certain choices because they are not adequately informed or because they would have a detrimental effect on others. All societies in fact provide some form of regulation as to acceptable private behaviour - the prohibition of the use of narcotics and restrictions on consumption of alcohol and tobacco are prime examples.

A CBA can be implemented at any of three stages of a project:

Pre-project appraisal - looking at a planned project (or projects) in advance.

On-going evaluation - looking at an operational but not completed project.

Post evaluation - looking at a completed project.

Evaluation Criteria

The gathering of data should be done in the context of what one needs to know in order to make wise decisions. This task is simplified if the streams of future costs and benefits can be brought together to one point in time, and expressed in a common unit, so far as possible. In CBA the most common evaluation criteria is the Net Present Value (NPV). This is defined as the present value of all benefits minus the present value of all costs[5].

However, *sustainability criteria* may be imposed on CBA as a framework for evaluation as well. The operational details of these are as yet problematic and need to be developed. A discussion of this topic lies outside the scope of this paper[6].

Income Distribution

Assessing *distributional consequences* is a part - implicitly or explicitly - of the use of evaluation criteria, since the net present value contains the sum of individual costs and benefits. The fact that each individual is generally given the same value, does not relieve us from the responsibility of considering this aspect. This appears particularly relevant in a developing country context. While the traditional focus of CBA is on income maximization as an aggregate measure, there are ethical positions that would focus on the lot of the very poorest (Rawls, 1971). The position taken here is that:

1. The search for any significant effects on income distribution should be mandatory in CBAs.
2. That any significant effects should be presented to the decision-makers.
3. That such effects should not be weighed implicitly, but that explicit values effecting the decision should be illustrated. It could be stated that: "assigning a double value to the benefits accruing to landless labourers would turn the project from a loss-making into a profit-making one."

Identification of Costs and Benefits

A common mistake in the appraisal situation is to limit unnecessarily the number of alternatives considered. Often, only a single project idea is presented. A thorough

CBA critically looks at a variety of available alternatives and may therefore result in a better end project.

Quantification of Costs and Benefits

This usually presents problems since our knowledge of the underlying natural scientific relationships is quite incomplete. This is a problem for all kinds of evaluation techniques, CBA or non-CBA. However, decisions still have to be made, and CBA can be a useful aid in organizing available information.

Valuation of Costs and Benefits

This entails applying "social price tags" to environmental impacts. The extent to which environmental effects can actually be expressed in monetary terms will differ depending on their nature and their relationship with existing markets. In some cases, one has to look for indirect market relationships, or even construct artificial markets to reveal people's valuation of environmental consequences (Freeman, 1979; Hufschmidt et al., 1983; Dixon et al., 1986). Valuation approaches are discussed at length later in the paper.

Discounting

Discounting uses a *real* social rate of discount. Inflation is a separate issue and need not concern us here. A general, uniform rise in the value of all costs and benefits will not affect our evaluation criteria.

Most people find it natural to regard the value of a dollar today as greater than the value of a dollar in ten years. This holds true even if the future dollar is guaranteed to be returned and with compensation for inflation. Several reasons could explain this position:

1. A dollar now could be invested and could therefore be worth more in ten years (there is an opportunity cost in terms of return on capital foregone).
2. A person could be richer in ten years, so an extra dollar will mean less then (the marginal utility of income will diminish).
3. One is impatient to use the dollar now rather than later (the pure rate of time preference).

These perceptions have been formalized by economists[7] and led to three major, but not distinctly separate, approaches in the determination of actual rates:

1. The *social opportunity cost of capital* approach, which looks for empirical evidence of (before tax) profits on alternative investment opportunities. The main argument here is that public investment displaces private investment with this return.
2. The *consumption rate of interest* approach, which is based on market data

revealing consumer preferences for consumption today versus tomorrow. Empirically, this entails looking at (after tax) returns to the investor on monetary instruments such as risk-free bonds.

3. The *social rate of time preference* approach. This takes the rate to be mainly a political parameter set on the basis of considerations related to the reasons for discounting discussed above: (a) the per capita income growth perspective, (b) the rate at which utility of increases in marginal income diminishes and, sometimes, (c) an assumption of the pure rate of time preference among consumers.

The arguments for different rates for different projects have been made (Price, 1973; Cooper, 1981). However, to make CBAs comparable, certain national standards are needed, one of them being a consistent social rate of discount.

Arguments have been made for adjusting the rate of discount both upward and downward to account for peculiarities in environmental analysis (Cooper, 1981; Brown, 1983; Prince, 1985). However, it appears odd to value impacts differently because the label "environmental" could - perhaps with great difficulty - be attached to them. These arguments are rejected in favour of explicit standards of sustainability.[8]

The Time Horizon

While some economists may be myopic, it is unfair to argue that economics as a science is necessarily so. In principle, an infinite length of time could be considered. In practice, CBAs often limit the horizon to 20, 30 or 50 years, because the effects of a project tend to diminish after time. The choice of time horizon is also related to the choice of discount rate. The higher the discount rate, the lower the weight attached to long-term effects.

Uncertainties and Risks

This point is problematic for all environmental evaluation, economic and non-economic, CBA or non-CBA. While economics does not contain any magic solution for the lack of perfect information, it does suggest how to deal with this situation.

Its main tool is "sensitivity analysis", where the sensitivity in the value of the evaluation criteria (usually NPV) is tested for different assumptions regarding prices, costs, productivity changes and so on. Of particular interest are "switching values": assumptions for which the value of NPV changes sign, thus implying a different decision. An example would be the statement that "costs could increase by 24 per cent before the NPV turns negative."

Adjustments for risk are best made when estimating specific probabilities for costs and benefits rather than through a general adjustment in the rate of discount. This is because the assignment of a "risk premium" (raising the rate of discount) implies the assumption of a geometrically growing risk. This may be a quite inappropriate modelling of actual risk development over time.

Policy Conclusions

Conclusions should be drawn in terms of the criteria used, considering the planning goals that the policy-maker has defined. Decision-makers may want to consider *criteria other* than economic efficiency (possibly with weighting for equity). Whether the analyst should bring such considerations directly to the CBA is a controversial matter. To the extent that such criteria have been explicitly stated, however, it appears that they ought to be discussed.

An interesting subject for policy analysis is the *difference between financial and economic profitability*. Assuming that the project has a positive economic net benefit, the financial NPV may have to be manipulated. For example, public subsidies to soil conservation may be necessary to induce upland farmers to protect a watershed for the benefit of irrigation and hydropower generation downstream. A tax on the discharge of polluted water may be necessary to signal to a chemical factory that its wastes are causing environmental damage to the fish downstream. In general, projects with a negative economic NPV should not be undertaken.

Some projects may have an overambitious element of subsidy built into them from the start. Instead, the provision of enough information may be sufficient for private households and companies to implement financially attractive and environmentally beneficial projects. For example, soil conservation may be profitable enough for the individual farmer given the right natural and financial environment, but he may not have sufficient technical knowledge, or understanding of the long-term effects of slow erosion. In the other example, treatment of waste water may recover material of significant commercial value for an industrial plant.

4.2 Valuation Approaches

Valuation is one of the more difficult components of CBA, not least because there exists a divergence between private (financial) and social (economic) perspectives and needs. This occurs for the following reasons:

1. The availability of *information*. Results from experiments and theoretical calculations are often not available for individuals, particularly in developing countries where illiteracy creates a barrier to the flow of information. This is not to say that "the experts know best." Much information is also stored locally among ordinary people.
2. The *identification* of relevant costs and benefits. The individual can be expected to care primarily or exclusively for the property (land, cattle, factory) he or she manages. For society, environmental impacts extending over these boundaries are also relevant.
3. The *valuation* of costs and benefits may also differ significantly. The individual will react to market prices, whereas society needs to consider the real social costs/benefits.

4. The *discount rate* may also differ. Commonly the private rate is assumed to be higher than the social, because of high private risk-aversion, particularly among the very poor.
5. The *time horizon* can vary quite a bit among individuals. It is commonly assumed that households take a more short-term view of investment and consumption. Thus, society must uphold the long-term perspective.
6. The perception of *risk* is likely to differ. Individuals may suffer crucial blows to their own finances in the face of environmental degradation, or if investments to counter degradation fail. Society can rely on spreading risk (taxing the whole population for losses) and risk-pooling (some projects work, others fail, but they often balance out).
7. As for *income distribution*, one can expect individuals to care primarily for their own households. The task of regulating income distribution at large is seen as an overriding social task.

Valuation from a social point of view can be grouped in three categories:[9] *conventional markets, implicit markets, and artificial markets.*

Valuation Using Conventional Markets

The use of *conventional* (ordinary) markets does not necessarily mean that financial market prices are adopted without alteration. When significant distortions prevent a free market from functioning, appropriate economic ("shadow") prices have to be estimated. These should reflect the consumers' actual willingness to pay.

Consider the cost of project *labour*. The financial cost is the market wage. The economic cost is the opportunity cost to the economy; what does not take place elsewhere due to the employment of this person. In many developing countries with high and chronic unemployment the opportunity cost may be zero. This is a simplification, since even an unemployed person produces something more informally, such as child care, food for household consumption and gathering of fuelwood. However, the simplification of using a zero shadow wage rate may be closer to reality than using the market wage. For unskilled labour the latter is often influenced by a minimum wage standard, reflecting distributional concerns rather than opportunity cost.

Another example of the use of shadow prices, which is also relevant for many developing countries, is the adjustment of the official *foreign exchange* rate. Often, the domestic currency is overvalued. This makes import cheap for the privileged few who control the foreign currency, and avoids the politically unpalatable devaluation of the domestic currency. At times, the official exchange rate is quite out of line with actual willingness to pay for foreign currency, resulting in black markets.

Four types of analysis fall under the heading of valuation using conventional markets: changes in production, replacement cost, preventive expenditure, and the human capital approach. Which approach to use depends on the problem under analysis. Sometimes several approaches can be combined: The emission of SO_2 may affect plant production negatively (approach 1), necessitate the replacement of corroded material (approach 2), cause the removal of sensitive objects (approach 3) and cause damage to human health (approach 4). However, care must be exercised in order not to double count for costs or benefits when using several approaches simultaneously.

Many environmental impacts have a direct bearing on *production values*; pollution may directly damage the production of fish, potable water or irrigation water, soil erosion may diminish the value of crops grown in an area, and overgrazing may cause cattle production to drop. The analysis of these effects is a straightforward extrapolation of traditional CBA, applying the same principles to environmental phenomena.

The *replacement cost* method describes the cost of repairing environmental damage such as that caused by erosion, flooding and corrosion. The impact of erosion may be (at least partially) countered by the use of fertilizer, flood damage may demand the reconstruction of a village, and corrosion will necessitate more frequent replacement of metal and painting. The cost of these items can all be calculated. Provided that replacement actually occurs, this is an environmental cost.

The *preventive expenditure* method analyses the costs individuals have incurred in trying to mitigate environmental hazards. Examples include improved ventilation against radon, three-glass windows or protective walls against noice, buying bottled water or cleaning devices to avoid polluted flood water, or the using protective clothing against agricultural chemicals. Such expenditures can be compared with the cost of cleaning communally available water or using other pest management methods.

The *human capital* method entails the calculation of income losses and health care costs as a result of environmental damage. This is not sufficient, but may give minimum estimates for environmental health damage. The inclusion of alternative measures, such as the psychological cost of suffering will have to be done subjectively. From a developing country perspective, it is of particular concern that human capital values reflect existing income distribution.

Valuation Using Implicit Markets

The basic idea behind this set of techniques is that there are links between the consumption of ordinary goods sold on the market and the consumption of non-marketed goods, which include environmental values. Thus, changes in environmental quality will also be reflected in prices of ordinary goods, such as land and houses. These implicit prices reflect willingness to pay for environmental quality and are sometimes referred to as "hedonic prices" (Freeman, 1979).

The same basic idea has also inspired the study of travel costs in relation to recreation, as a means to capture willingness to pay for implicit environmental values. Originating with the work of Clawson and Knetsch (1966), this method has become widely used in the industrialized world, especially in the USA. Third World applications are few but do exist (Dixon & Hufschmidt, 1986).

There are major problems in applying these techniques in a Third World context. *Lack of market transactions and market data* is a considerable obstacle. In some cultures, land may not be (officially) sold at all, which prevents a market valuation of land degradation. If houses are not bought and sold on an active market, there is no basis for registering how environmental differences affect prices. More market-oriented policies in developing countries may make this approach more relevant in the future.

Valuation Using Artificial Markets

It is not always possible to make inferences from actual behaviour as in the approaches presented above, when identifying private valuations. Instead, consumer preferences may have to be measured by creating hypothetical situations using *artificial* markets.

This is done by first describing the environmental choices to consumers as succinctly and objectively as possible. They are then asked to express their willingness to pay for an environmental improvement or to prevent an environmental deterioration. Alternatively, the respondents could be asked to express their willingness to accept compensation. Often, this is done by asking questions in an iterative fashion, starting with an initial bid, and adjusting it according to the response (bidding game).

The strength of these methods is that they can be applied to a variety of situations where no other data is available. Artificial markets have been used to test individual valuation regarding water and air quality, aesthetic beauty, recreational values, preservation of open farmland, natural environments, disposal of hazardous waste, risk in relation to air travel, car travel, cigarette smoking and nuclear energy.

There are many difficulties pertaining to the use of artificial markets. However, some of these can be statistically tested for in retrospect, and to some extent controlled through careful survey design. The appropriateness of these techniques will vary, depending on the purpose of the study. They should at least be considered and carefully assessed in cases where individuals' willingness to pay cannot be registered in an existing market.

For subsistence economies it may be inappropriate to survey the willingness to pay. A more appropriate way of expressing the same logic would be to use "willingness to donate labour" for some environmental cause, say tree-planting or the construction of water diversions.

4.3 "Alternatives " to CBA

The appropriateness of a particular method of economic evaluation depends on the task at hand. The framework presented above is very flexible and can easily accommodate methods that are sometimes presented as "alternatives".

An *Environmental Impact Assessment* (Clark et al., 1978) can be seen as a complement, even as a necessary foundation for a proper CBA. The EIA specifies the impacts for the CBA to evaluate.

Cost-effectiveness Analysis (CEA) is a useful approach when it is impossible to estimate benefits. Certain environmental goals may have their basis in political targets: "The amount of pesticides in agriculture should be reduced by 30 per cent in ten years", or: "Tree planting must increase so that the mean annual increment of fuelwood species equals the fuelwood demand in year 2000." The decision-makers want the least cost method of achieving this goal. The assumption - which may or may not be valid - is that the least cost is less than the value of the unquantified benefits.

The opportunity cost approach (Hufschmidt et al., 1983) measures what has to be given up in order to preserve a non-priced asset. If a part of the tropical rain forest is declared a reserve, how much more does it cost to provide agricultural land for people elsewhere? By comparing the costs of development at different sites, a trade-off could be made with anticipated environmental effects.

Objective contribution matrices (Birgegård, 1975) is a device for selecting alternative projects on the basis of their contribution to multiple social goals. This does not stand in opposition to the use of CBA, as the contribution to goals other than economic growth such as income distribution, employment creation and regional balance can be easily added to the CBA framework as weights or restrictions.

The Choice of a Method

Rather than entering into a detailed discussion of CBA versus other approaches, it is suggested here that available alternatives could be screened according to a number of criteria such as:

1. Operational cost - should not be greater than the value of the information produced.
2. Comprehensiveness - the degree to which the approach takes relevant impacts into account.
3. Comprehensibility - the degree to which the method and its results can be usefully interpreted by the users of the information.
4. Democracy - the degree to which the preferences of individuals are reflected.
5. Relevance for decision-making - the degree to which the information is useful to policy makers.

Criticism of CBA

There are critics who reject the foundations of CBA on philosophical grounds (e.g. Schumacher, 1973). Others are critical of its practical usefulness in dealing with environmental matters. However, the critique against the fundamental method of cost-benefit comparisons is often based on misunderstanding of what CBA aims to do. It is sometimes suggested that CBA is a form of "monetary reductionism" that claims to use "correct" prices (Söderbaum, 1987). Therefore it has been made explicitly clear here that CBA rests on certain basic assumptions.

The critique regarding the limited usefulness of CBA in *environmental* matters has some merits. Data are usually deficient, and sometimes overriding concerns about the preservation of a particular ecosystem will dominate the decision-making. Complementary principles of sustainability need to be developed (Markandya & Pearce, 1988). However, the critique regarding the limited availability of information strikes equally hard at *all* methods of evaluation, CBA or non-CBA. This is why one quotes with sympathy the pragmatic attitude of Maurice Chevalier, who responded to the question of how he perceived old age with the following remark: "Well there is quite a lot wrong with it, but it isn't so bad when you consider the alternative."

There are numerous applications of the methods discussed here. Easily accessible examples are found in Dixon and Hufschmidt (1986), Dixon et al. (1988), Bojö et al. (1990), Dixon et al. (1990) and Stocking et al. (1990). These show that CBA of environmental impacts is not merely a theoretical construct. It can be usefully applied to real life environmental problems, in developing as well as developed countries.

ACKNOWLEDGEMENTS

The author would like to gratefully acknowledge constructive comments from Thomas Andersson, Hans Bergman, John A. Dixon, Tomasz Zylicz, and several participants in the Advanced Seminar of the Department of International Economics and Geography at the Stockholm School of Economics.

NOTES

1. The term "neoclassical" refers to a school of thought within economics that has its roots in late 19th century theory as developed by economists such as Walras, Jevons, Menger and Marshall. The prefix "neo" serves to distinguish this school from the earlier "classical" economics with Smith, Ricardo and Mill as the leading theorists. For details, see Landreth and Colander (1989).

2. The term "minimize" is avoided. The minimum environmental damage possible is none at all. However, this is incompatible with human civilization, at least above the Stone Age level. Thus, the level of environmental damage should be reduced to the point where marginal benefits equal marginal cost. This point may be significantly different from the zero level.

3. In fact, several important environmental problems are international concerns: the "greenhouse effect", depletion of the ozone layer, acid rain and so on. However, these issues lie outside the scope of this paper.

4. There is a vast literature on CBA. Works of major importance include Dasgupta et al. (1972) and Little and Mirrlees (1974). A good and easily accessible introduction is Gittinger (1982).

5. In formal terms:

$$NPV = \sum_{t=1}^{T} (B_t - C_t)/(1 + r)^t$$

where t = time index in years; T = time horizon in years; B = social benefits; C = social costs; r = social rate of discount.

6. There is an extensive and rapidly growing literature on "sustainability". Useful references are Barbier (1987), Goodland and Ledec (1987), WCED (1987), SGN (1988), and Turner (1988).

7. See Lind (1982) and Markandya and Pearce (1988) for good reviews.

8. See Markandya and Pearce (1988) for an extended discussion.

9. The terminology is from Bojö et al. (1990). This source in turn draws on Hufschmidt et al. (1983) and Dixon et al. (1988).

REFERENCES

Barbier, E. 1987. The Concept of Sustainable Economic Development. *Environmental Conservation* 14:101-110.

Bojö, J., Mäler, K.-G. and Unemo, L. 1990. Environment and Development: An Economic Approach. Economy & Environment Series No. 2. Kluwer Academic Publishers, Dordrecht, The Netherlands.

Birgegård, L.-E. 1975. The Project Selection Process in Developing Countries. Stockholm School of Economics, Stockholm.

Brown, S.P.A. 1983. A Note on Environmental Risk and the Rate of Discount. *Journal of Environmental Economics and Management* 10:282-286.

Clark, B.D., Chapman, K., Bisset, R. and Wathern, P. 1978. Environmental Impact Assessment in the USA: A Critical Review. Department of the Environment, Marsham St. 2, London.

Clawson, M. and Knetsch, J.L. 1966. Economics of Outdoor Recreation. The Johns Hopkins University Press, Baltimore.

Costanza, R. 1984. Natural Resource Valuation and Management: Toward an Ecolgical Economics. In: Jansson, A.M. (ed.). Integration of Economy and Ecology: An Outlook for the Eighties. Proceedings from the Wallenberg Symposia. Askö Laboratory, Stockholm University, Stockholm. pp. 7-18.

Cooper, C. 1981. Economic Evaluation and the Environment. Hodder and Stoughton, London.

Dasgupta, P., Marglin, S. and Sen, A. 1972. Guidelines for Project Evaluation. UNIDO. United Nations, New York.

Dixon, J.A. and Hufschmidt, M.M. (eds.). 1986. Economic Valuation Techniques for the Environment: A Case Study Workbook. Johns Hopkins University Press, Baltimore.

Dixon, J.A., James, D.E. and Sherman, P.B. (eds.). 1990. Dryland Management: Economic Case Studies. Volume II: Case Study Reader. Earthscan, London.

Dixon, J.A., Carpenter, R.A., Fallon, L.A., Sherman, P.B. and Manopimoke, S. 1988. Economic Analysis of the Environmental Impacts of Development Projects. Earthscan, London.

Freeman, M. 1979. The Benefits of Environmental Improvement. Resources for the Future, Johns Hopkins University Press, Baltimore.

Gittinger, J.P. 1982. Economic Analysis of Agricultural Projects. Second Edition. Johns Hopkins University Press, Baltimore.

Goodland, R. and Ledec, G. 1987. Neoclassical Economics and Principles of Sustainable Development. *Ecological Modelling* 38:9-18.

Helmers, F.L.C.H. 1979. Project Planning and Income Distribution. Martinus Nijhoff Publishing, The Hague.

Hufschmidt, M.M., James, D.E., Meister, A.D., Bower, B.T. and Dixon, J.A. 1983. Environment, Natural Systems and Development: An Economic Valuation Guide. Johns Hopkins University Press, Baltimore.

Landreth, H. and Colander, D.C. 1989. History of Economic Theory. Second Edition. Houghton Mifflin Co, Boston.

Lind, R.C. 1982. Discounting for Time and Risk in Energy Policy. Resources for the Future/Johns Hopkins University Press, Baltimore.

Little, I.M.D. and Mirrlees, J.A. 1974. Project Appraisal and Planning for Developing Countries. Heinemann, London.

Markandya, A. and Pearce, D. 1988. Environmental Considerations and the Choice of the Discount Rate in Developing Countries. Environment Department Working Paper No. 3. World Bank, Washington D.C.

Odum, H.T. 1984. Embodied Energy, Foreign Trade and Welfare of Nations. In: Jansson, A.M. (ed.). Integration of Economy and Ecology: An Outlook for the Eighties. Proceedings from the Wallenberg Symposia. Askö Laboratory, Stockholm University, Stockholm. pp. 185-199.

Price, C. 1973. To the Future: With Indifference or Concern? - The Social Discount Rate and its Implications in Land Use. *Journal of Agricultural Economics* 24:393-398.

Prince, R. 1985. A Note on Environmental Risk and the Rate of Discount: Comment. *Journal of Environmental Economics and Management* 12:179-180.

Rawls, J. 1971. A Theory of Justice. Harvard University Press, Cambridge.

Schumacher, E.F. 1973. Small is Beautiful: A Study of Economics as if People Mattered. Sphere Books Ltd, London.

SGN, Stockholm Group for Studies on Natural Resources Management. 1988. Perspectives of Sustainable Development: Some Critical Issues Related to the Brundtland Report. Stockholm Studies in Natural Resource Management No. 1. Department of Systems Ecology, Division of Natural Resources Management, Stockholm University, Stockholm.

Stocking, M., Bojö, J. and Abel, N. 1990. Economics of Agroforestry. CAB International, Wallingford.

Söderbaum, P. 1987. Environmental Management: A Non-Traditional Approach. *Journal of Economic Issues* 20:139-165.

Turner, R.K. (ed.). 1988. Sustainable Environmental Management: Principles and Practice. Belhaven Press, London.

WCED, World Commission for Environment and Development. 1987. Our Common Future. Oxford University Press, Oxford.

Linking the Natural Environment and the Economy; Essays from the Eco-Eco Group,
Carl Folke and Tomas Kåberger (editors)
Second Edition.
1992. Kluwer Academic Publishers

CHAPTER 4

Measuring Instrumental Value in Energy Terms

by

Tomas Kåberger

Physical Resource Theory Group
Chalmers University of Technology
S-412 96 Gothenburg, Sweden

Energy is a common physical measure in both economic and ecological systems. When attempts are made to integrate economic and ecological perspectives of systems energy terms are sometimes used. In this chapter I attempt to compare methods using historical measures (embodied energies) and measures of present energy. It is concluded that these measures are very useful, are perhaps unnecessary, and are certainly not sufficient for the design of an environmentally acceptable economic theory or policy.

1. INTRODUCTION

Using energy as a common numeraire is an attractive possibility in the analysis of interacting ecological and economic systems. There have been descriptions of economic systems using energy-accounting for more than a hundred years (Cleveland, 1987; Martinez-Alier, 1987). Even earlier the physiocrats emphasized the importance of physical dimensions of economic analysis. A revival of attention to the energy approach as an alternative to monetary analysis appeared during the present century with the Technocratic movement. This movement reached its greatest popularity in the US in the period between the world wars, and this has has been described by Atkin (1977). Taylor (1988) describes the connection between the ideas of the technocrats and the post World War II developments of systems ecology led by Howard Thomas

Odum. According to Berndt (1985), the technocratic ideology also formed the basis for the energy analysts of the 1970's. Articles on the methods and issues of the 1970's energy analysis of economic systems were collected by Thomas (1977) and Gilliland (1978). A review of the results of energy analysis was made by Spreng (1988).

Ecological energetics of this century can be traced back to Alfred J. Lotka. In his book Elements of Physical Biology published in 1924 (reprinted 1956 as Elements of Mathematical Biology), he attempted to describe complex ecological systems using terms and quantities from thermodynamics. Measuring energy flows in real ecosystems requires, however, substantial research resources. Lindeman (1942) started, and he was followed by H.T. Odum during the 1950's, who then led energy flow studies on a number on different ecosystems. Slobodkin (1960) seems to be the source of general statements on efficiencies in the energy transformations between trophic levels in the food chain.

Present work on analysing the role of the human economy as a part of the ecological system has been carried out mainly by ecologists inspired by H.T. Odum. His popular book "Environment, Power and Society" was published in 1971 and together with Odum & Odum (1976) it has formed the basis for many ecologists attempting to describe the interaction between society and nature in energy terms.

1971 was also the year of Georgescu-Roegen's book "Entropy and the Economic Process" which is more advanced from a thermodynamic point of view. Not only does he deal with the flows of conserved energy, he also considers the importance of the irreversible production of entropy.

Diagrams showing flows of energy have been used to describe different complex systems of interacting units. Energy flows in engineered processes are often described in so-called Sankey-diagrams. Systematic techniques for making diagrams of complex systems have also been developed in electrical engineering. According to Taylor (1988), Odum first introduced diagrams inspired by electrical circuit diagrams into ecology in his an article of 1956. These symbolic analogies, however, had implications beyond those intended, and Odum later developed the novel set of symbols and rules that he describes in his 1971 book, and in more detail in his 1972 article.

As interactions in ecological and economic systems are always coupled to flows of energy, these diagrams become informative, though simplified, descriptions of interaction patterns. Odum's diagram techniques have been used to produce a very information-rich picture showing the quantitative structure and interactions of the natural and the man-made world on a regional level (Jansson & Zucchetto, 1978).

The engineering-based energy analysis, Odum-type energy analysis of society in the environment, and the thermodynamic analysis of the economic processes inspired by Georgescu-Roegen are some of the irregular corner stones on which attempts are made to build ecological economics.

Descriptions of economic systems based on monetary measures leave out real transactions that are not coupled to the transactions of monetarized values. When the same economic system is described in terms of energy flows, real transactions that are omitted in the monetary description, are revealed. Many of the real effects that are not marketed and payed for are connected with environmental destruction. Free resources taken from nature, and pollution are rarely seen in the economic description but they appear when the same system is mapped using energy terms.

This article deals with the usefulness of energy measures when assessing the values of different objects or alternative actions. There are instrumental values, or resource values, that arise because the "valuable" object or action is useful in order to reach a given aim. There are also final values, or inherent values, ascribed to entities irrespective of other purposes or consequences. Environmental controversies typically involve both types of values.

2. THE LAWS OF THERMODYNAMICS

2.1 Conservation

The *first law* of thermodynamics tells us that energy is a conserved scalar quantity. Any change in the energy content of a system is due to flows of energy across the system boundary.

When looking at systems that are only partly understood, this conservation law makes energy a very useful quantity. Even without knowing what processes go on, one may use energy balance equations to derive unmeasurable quantities. For this reason, energy has long been the most important quantity in the analysis of physical and chemical systems, and this is also the reason why it is used in ecology and economics.

2.2 Consumption

Energy does not measure ability to perform work. While energy is conserved in any process, its ability to drive physical processes is irreversibly lost as entropy is produced. According to the *second law* of thermodynamics, this is the case for all real processes.

The amount of energy that can be transformed into work is called the exergy. The exergy concept was introduced by Willard Gibbs (1873), the term exergy was suggested by Z. Rant (1953), and a general definition was given by Baehr (1965).

Exergy is consumed in all real processes. When we look for resources to drive engines, feed people, or produce electricity we look for exergy. Energy as ambient heat is of no help, while fuels in an atmosphere with oxygen, solar radiation reaching our relatively cold Earth or water with potential energy above ocean level are useful physical resources.

3. MEASURES OF ENERGY

Several kinds of energy quantities may be used for analysing systems which convert energy and materials or the products of such systems. In the following section I divide them into *measures of present energy* based on the actual energy content of a component, and *measures of historical energy costs* that depend on how the component was produced. The first set of measures in turn is subdivided according to whether the energy measure is a first law energy measure or whether it takes the second law into account. The historical measures are divided into specific forms of primary energy sources.

3.1 Measures of present energy

One set of quantities we may call measures of *energy content*. Often, but not always, energy content is the same as enthalpy of combustion. The energy content of a barrel of oil and the energy contents of foodstuffs are normally defined as the heat released in processes when these substances are burned.

To be precise, the energy content must be carefully specified regarding the state of the surroundings: What final state is considered an equilibrium state and what processes are available? Considering oil and foodstuffs we may find the specifications obvious, but such specifications are clearly necessary when dealing with fissile material brought into a nuclear reactor: What fraction of the mass is available for conversion into heat? Should the energy of neutrinos be included?

A similar set of quantities we may call figures of *exergy content*. Here, measures are aimed at quantifying the maximum work available as the component is brought into equilibrium with its surroundings. As with energy content measures, both the equilibrium state and the available processes must be specified.

Several definitions of standard environments and available processes are possible. The energy of water in a dam above a hydro-electric plant may be defined as the energy that can be extracted in that particular plant, it may be the potential energy of the water with respect to the water level below the plant or with respect to sea level, but the energy may also include the nuclear energy available if the nucleons of oxygen and hydrogen atoms were rearranged by nuclear reactions into nuclei of iron.

These two sets of energy quantities depend only on the state of the subsystem under consideration in a certain environment. Position in time, or the history of the system has no relevance. One can imagine using a device, performing the defined available processes, to actually measure the content of energy or exergy of the system in its surroundings. No information is required about the roles of these entities in a larger system, nor about the way they were produced, not even whether it was arficial or natural processes, which were involved.

The difference between these two sets of measures is important. First of all: One joule of a particular form of energy is not necessarily transformable into one joule of another form of energy. Exergy is consumed in all economic processes and cannot be reproduced, and as a consequence it is often misleading to compare a joule of exergy-poor heat with a joule of electricity. As mentioned above, this is a consequence of thermodynamics which was most forcefully brought into economics by Georgescu-Roegen (1971).

This difference in usefulness, or resource value, of a joule of energy has been recognized and systematically used by Wall (1986) resulting in informative descriptions of different industrial societal systems (Wall, 1987, 1990).

Exergy flow diagrams illustrate how energy flows in society are subject to irreversible entropy production. As entropy is produced, exergy is lost and the energy becomes less useful. In energy terms there is no difference between the potential energy of water above a hydroelectric power plant, the energy of hot steam in an industry, or the waste heat given back to nature. But for economic purposes the usefulness of these flows is very different. The hot steam or waste heat cannot be fully used to produce electricity or motion while the potential energy of water can. This illustrates why electricity is generally more useful than steam, and why we bother to build dams for hydro-power while we pour waste heat into nature, and ultimately into space. Approximate exergy contents of various forms of energy are given in Table 1.

Table 1. Exergy content of various forms of energy

Type of Energy	Exergy as a percentage of the energy
Potential energy	100
Electricity	100
Chemical fuels	100 (approximately)
Solar radiation	95 (Karlsson, 1982)
Hot steam	50
Space heating	10-20
Earth radiation	0

Odum uses energy measures, and collects arguments supporting the relevance of his methods for ecosystem analysis. He does not use the second law of thermodynamics, in a physical sense. It is, however, interesting to note that Odum, while claiming to map energy flows, is in fact close to mapping flows of exergy, as long as he deals solely with ecosystems. The kinds of energy that goes into conversion processes in ecosystems are solar radiation and the chemically bound energies in various forms of biomass. In his diagrams heat flows are drawn in analogy with grounding flows in electrical circuits. Thus the problems of irreversible processes never appear in the descriptions. Since both solar radiation and biomass have exergy to energy ratios close to one, and the consideration of exergy-poor heat flows is removed by his methodology, differences from exergy accounting are small.

When turning to economic systems, however, Odum's methodology gives a very different quantitative description from the methods of Wall because of the problem of comparing energies with different exergy contents.

Using thermodynamics, it is in theory possible to obtain exergy contents of materials such as ores and refined materials in circulation in an ecosystem or a society. In practice, however, such estimations have turned out to be very difficult to carry out, especially for mixtures of different substances.

The actual measurement of energy content is carried out using a bomb calorimeter. This gives a measure of the energy released as heat when a fuel burns. In practice, the results of such measurements are adjusted using models of some unintended processes such as heat losses to the surroundings.

Measurements of exergy content are typically much more complicated since in this case compensation must be made for all the irreverible processes using some sort of model of the measurement process. Avoiding irreversibility often requires infinitely slow processes, and if you want the measurement result you must introduce irreversible processes. Indirect measurements and calculations of exergy content using a model of the system, are therefore the typical method of acquiring exergy content measures. We cannot measure systems that we cannot model appropriately.

Yet, the simple principle that exergy is consumed as concentration gradients decrease is useful for understanding some processes important for the interaction between the economic and ecological systems. Ayres (1978) followed by Chapman and Roberts (1983) describe the process whereby metals dispersed in the Earth's crust are concentrated by human industrial systems using exergy. Of the exergy added to the process, most is consumed, while some is stored in the product such as the metallic aluminium produced from bauxite.

When metal is in use in a society some of it is once again dispersed. It does not go back into the Earths crust, but into the biosphere as pollution. As the metal atoms are spread out the exergy stored is irretrievably lost.

From practical experience we know that the exergy which is required in a process to reconcentrate the atoms dispersed into the biosphere is typically very large. Despite bioaccumulation, it is not even feasible to extract mercury from pike in polluted lakes. It is also clear that household wastes contain materials that are valuable resources as long as they are not mixed with each other. However, a sensible method of quantitative thermodynamic modelling at this level remains to be found. At the scale of mixing bottles and newspapers thermodynamic modelling seems to serve only as an analogy. It is by technological and economic arguments we have learned that unless metal objects are recycled befored the metal is dissipated into the biosphere, the exergy requirement is less when starting from virgin ores.

Measures of present energy may, in theory, be used as measures of resource values when the aim is to produce mechanical work or heat. As described above, application is made difficult by the problems of designing realistic measurement processes or models.

Economic market evaluations will generally show what fraction of the theoretical resource value can be utilized using available technology. This may often be the most relevant evaluation method. However, an analysis based on contents of energy or exergy can show where it might be fruitful to attempt to develop technology. For example, there is normally a difference in economic value per unit energy of chemical fuels and electricity. With present technologies car engines, gas turbines or steam turbines, only up to around 50 per cent of the exergy present in the fuels going into the process can be found as work. This may be contrasted with present electric motors which have efficiencies of 80-95 per cent.

While measures of present energy may give information on instrumental values, they provide no information on the final values involved. It is, hopefully, obvious that present energy measures are irrelevant when deciding whether to save a chair or a child from a burning house.

3.2 Historical measures.

Once the structure of the system in terms of energy flows is known, it is possible to investigate the total amount of energy used in the production process of a certain subsystem or object.

The non-trivial feature of this historical energy measure is that it contains information not only on the energy contents of the inputs to the final production process, but also on the energy spent in producing these inputs. The energy costs of inputs are traced back until the energies accounted for constitute the energy sources (or primary energies) of the system. Considering a natural ecosystem, solar radiation is the main source of energy, while an industrial system may have fossil fuel as the primary source of energy.

The result of such an investigation is a total *energy cost of production*. Several processes, using different amounts of energy, may result in identical entities. Thus energy cost of production is dependent on the history of an object, and measurements on the object itself are not enough to obtain the energy cost of production. Often little, or nothing, of the primary energy used in the production process is incorporated in energy content of the final product.

A common term used for the energy cost of production is *embodied energy*. Odum also uses the abbreviated form *emergy* (energy memory)(Odum, 1984). As the quantities of energy that have been added to yield the energy cost of a production measure are only partly present in the animal or product under consideration, the word "embodied" may be misleading.

The importance of determining the energy cost of production measures is evident from the following example: Compare two industries with identical production in a country with fuel-based electricity production. Industry A annually uses 5 GWh of electricity together with 15 GWh of fuel, while industry B uses 10 GWh of electricity and 5 GWh of fuel. Looking only at the energy content of the inputs, you may conclude that industry B is more energy efficient since the energy content of its inputs is only 15 GWh compared to 20 GWh for the more fuel consuming industry A. Considering the historical costs of the energy carriers entering the factories, however, leads you to the opposite conclusion. The electricity is produced from fuel further down the system and for each GWh of electricity 3 GWh of fuel is consumed, thus making the total fuel cost for running the industries 30 GWh and 35 GWh respectively. Thus, decreasing the production in industry B would save more fuel then doing the same in industry A.

This analysis of the energy cost of production gives a clear result as to which of the processes is the most efficient since fuel is the only ultimate source of energy. The historical measure is in fact *the fuel cost of production*.

In ecological systems the ultimate energy source is normally solar radiation. Thus, it is possible to calculate *the solar energy cost of production*.

Once the ecological systems are used for human purposes the situation changes: Apart from solar energy there are often fossil fuel energies added to increase production. The solar energy cost of production is then a measure of areal efficiency. Fossil fuel cost of production measures appears as informative indicators of sustainability. Investigating the different fuel and solar energy costs of production may be a tool to compare different management strategies for natural resources, such as aquaculture (Folke & Kautsky, 1989).

The nuclear energy used in nuclear power plants has not been produced from solar radiation. The natural gas and oil found in the earth's crust seem to be the result of solar energy, but we do not know how much was required.

Thus, for subsidized ecosystems, with more than one primary source of energy, there is no unique way of comparing the different forms of energy as a single energy term, and there is no single criterion for energy efficiency. Nor can energy analysis result in any criterion on desirability, since we need input of final values affected by the alternatives.

Another difficulty with measures of energy cost of production lies in the problem of how to treat genetic information or human know-how. If the size of a population is drastically changed, this will have genetic consequences. Attempts to assess such costs in energy terms raises the question of the energy cost of evolutionary genetic development: What is the energy spent in the evolutionary process developing this particular salmon which is about to become extinct?

Depending on one's assumptions and system boundaries, one may correctly answer anything from a fraction of a joule up to the total energy of the universe. When evaluating alternatives in an environmental management issue, all these answers appear irrelevant so long as we are unable to use energy to reproduce the evolutionary process.

The problem of assessing the energy cost of the production of human know-how is related to the problems with genetic evolution. Inventions cannot be reproduced, and their energy cost can hardly be assessed.

The absence of a unique way of comparing different energy sources, and the presence of seemingly open-ended questions on how to treat the production of genetic information and human knowledge are two important limitations on the use of historical energy measures in the analysis of ecological and economic systems.

It has been suggested that there is a linkage between the energy cost of production, and the societal value of the object considered. Costanza (1980) showed that not only was there a correlation between the energy cost of production for a product and its market price, but that it was also possible to improve this correlation if an energy value was assigned to other economic costs of production.

It is not surprising to that this correlation is found between the economic cost and the energy cost when a large part of the economic production cost represents payment for the energy used. Nor is it surprising that the correlation can be improved by assigning increased energy equivalents to other factors in the production when these dominate the economic cost of production. When considering energy measures for evaluating economic alternatives it is important, however, to note that it is not the cost, but how much more is gained, that constitutes the societal value of consumption.

Despite the absence of transforming processes available for study, Odum (1988) has also assigned solar energy costs of production measures to other energy sources in addition to what he calls "human services." Using his solar energy cost of production measures, he can calculate a single figure even for processes where human labour and a mixture of energy sources are involved. This figure he calls "Solar EMERGY", spelled with capital letters in his 1988 Science article. He concludes: "Energy is not a measure of value, because the highest valued processes such as human services and information have tiny energies, but EMERGY does measure these values appropriately. (...) Thus EMERGY is suggested as the measure of contributions to the general economy, a means for evaluating environment , resources, and public policy alternatives" (Odum, 1988 p. 1136).

What Costanza and Odum have done is to go beyond the limit of what constitutes energy costs of production as an account of a physical process available for observation. In an attempt to reach conclusive results they offer other energy cost evaluation methods. They are then able, in a similar way as most economists do, to present the result of a complex analysis with a single number. Many of the critical arguments voiced by Söderbaum (Chapter 2 this book) against main-stream economists will also apply to anyone who would interpret above citation of Odum as justifying the use of these figures as inputs for simple decision criteria.

In a market which is functioning well, an approximate value for the economic cost or benefit of a change will be given by the market value of the consumer products supplied or withdrawn provided that the quantities are relatively small. In such cases, due to the correlation of market price and energy cost of production, the latter may well be a reasonable alternative to economic measures.

The societal value of what is produced is not the cost of production. The well-being generated, the consumer surplus, is something different. Consider the following examples:

1. Following the restoration of a building the desired indoor climate is produced using only 10 per cent of the energy used previously. The energy cost of restoration was very small as were the economic costs. The value of the indoor climate has not decreased, while the energy cost of producing it has.
2. The development of new technology for making metalic aluminium out of aluminium oxide reduced the electricity required to a mere 20 per cent of the 19th century requirements. The societal value of the metallic aluminium produced for a certain application has not.

The relationship between the energy costs of sustaining a society and the economic well-being of that society has an important role in debates on energy policy. It is possible to show that for a long time there has been a correlation between the standard of living and the use of commercial energy (Classen & Girifalco, 1986 is a recent example). Some have used this as an argument towards the conclusion that there is causality one way or the other. However, whole countries may increase their economic activity while lowering energy intake. An example is Japan: between 1973 and 1985 the per capita GNP increased by 46 per cent while commercial energy use decreased by 6 per cent, also per capita (Goldemberg et al., 1988, p.5).

In ecosystems the solar energy cost of production as a valuation criterion runs into similar problems. If a step in an evolutionary process includes the increased fitness of a species by improved digestion, this would decrease the solar energy required by every individual in the new population. It is, however, in contradiction with our ideas as to what constitutes values in ecosystems to conclude that the individuals in the previous, inefficient population were more valuable.

We know that evolution in ecological systems may improve efficiencies, and restoring a building may decrease heating requirements without any values being lost. This observation limits the use of energy cost of production as a measure of value.

In a vigorous critique of the application of energy cost of production measures in economic analysis, Webb and Pearce (1975) argue that these measures cannot compare different forms of primary energy, nor relate energy to other factors of production such as labour, and they conclude that the methods of energy analysis are useless.

The ideas of energy analysts that identifying the energy cost of production of different products in order to suggest changes in consumption patterns and thereby reduce the environmental costs as well as resource depletion is, according to Webb and

Pearce, superfluous. Imposing an energy tax would be easier and it would automatically have the desired effect on energy-intensive products.

However, in most countries the markets for electricity and fossil fuels are often characterised by monopolies, secret long-term contracts and government subsidies making the energy markets far from ideal. In Sweden, the most energy-intensive industries have their energy taxes reimbursed. As a result market imperfections are often so many that a single correction such as an energy tax, imposed for environmental reasons is insufficient to come closer to the desired situation. In real economies, energy analysis may then contribute additional important information.

Another situation in which energy cost measures of production are of interest are technologies in their research and development stages, where they are not yet subject to market evaluation. A thermodynamic analysis is an obvious step for any scientist intending to develop a successful technology. It is, however, not an obvious step for someone who aims at a long-term research grant. For this reason, in 1974 a law was passed in the United States intended to ensure that energy projects have the potential to deliver more energy than they require (Public Law 93-577, Non-nuclear energy research and development act, section 5).

4. INSTRUMENTAL AND FINAL VALUES

When describing the process of selection of what is good from what is bad it is useful to distinguish the problems of selecting the final aims from the problems arising when looking for the efficient methods of achieving these aims. The values relevant for the first category are final values, while the second category of problems are solved by analysing to what degree different options are instrumental in achieving the final aims.

Measures of present energy are measures of physical resources. If we had technologies available performing processes with arbitrarily small losses, exergy content would be the most appropriate measure of instrumental value for societal physical purposes.

In the real world, only a limited set of technologies are available with various efficiencies. The value of a resource then depends not only on the exergy content of the resource, but also on the availability of a technology for transforming the resource into the desired product or service.

The exergy content, and the resource value taking available technologies into account, are examples of instrumental values. Given the purpose of the exercise it measures the degree to which the resource may help to full fill that purpose. Such measures have nothing to say on the question as to whether the purpose is good or bad. Nor can these measures help you when choosing between competing uses for the resource.

The historical energy cost of production measures is useful in assessing which of several available processes most efficiently produces a desired result. It is a way of comparing the efficacies of the different technologies available for transforming available resources into a desired product or service. Thus you may see energy cost of production not as a measure of the product, but as a measure of the instrumental value of a process.

With a given stock of energy resources and a given final aim the system with the lowest energy cost of production has the greatest instrumental value for producing a final value from the resources.

The energy measures described above are useful in describing entities available as resources for consumption. They may be used to deduce the physical limits of the processes which it is possible to carry out, and they also serve the purpose of evaluating resource costs of real production processes. None of these measures, however, can tell you the final value of seeing, owning or destroying the entities involved.

Main stream economists will normally ascribe a final value only in accordance with human revealed preferences. In most applications of this principle to environmental issues a weak point is lack of knowledge. Of course, incomplete knowledge affects the human preferences. The systems analysis that may be carried out using energy measures may increase the knowledge about the system. Typically, the interdependence between species as well as chemical processes on different scales will be discovered. Disclosure of such interdependences, such as life-support functions of the natural environment described by Folke (Chapter 5 this book), will improve the relevance of the results of economic evaluation theories.

Remaining beyond the realm of energetics are the issues of final values inherent in other beings apart from humans. Assessments of final values of the existence, sufferings or utilities of other species are relevant to most issues of environmental pollution as well as to the issue of the exploitation of natural ecosystems. They are dealt with by Næss (1973), and Skolimowski (1981), on a theoretical level but they have never, to my knowledge, been explicitly introduced into applied environmental economics.

A second matter, central to most environmental issues and beyond what energy measures may be used for, concerns issues involving human rights or other hierarchical value systems. The question - Should the nuclear fuel cycle be allowed to kill 500 people in order to produce electricity and make the life easier for 1 million other people? - can not be answered by any measures of the energies involved. A logical method of analysing these issues is thoroughly presented by Rawls (1971).

The common types of energy measures described above serve as measures of instrumental values for physical resources and societal processes. The analysis necessary to obtain the measures may give important knowledge about the effects of the economic activities under consideration, thus giving important input to the decision process. However, as we have seen there are at least two kinds of issues involving final values that cannot in any way be resolved using energy measures.

5. CONCLUSIONS

Energy measures are useful for identifying, mapping and evaluating instrumental values in ecological and economic systems. The results from analysis of these systems using energy terms may provide an input to traditional economic methods of evaluation. The controversies arising because of the unsatifactory methods in economics to identify final values can not be solved by energy analysis.

The conclusion of the author is that using various energy measures in the analysis of interacting economic and ecological systems is very useful, is perhaps unnecessary, and is certainly not sufficient in order to design an environmentally acceptable economic theory or policy.

REFERENCES

Atkin, W.E. 1977. Technocracy and the American Dream: The Technocracy Movement 1900-1941. University of California Press, Berkley, CA.

Ayres, R.U. 1978. Resources, Environment and Economics: Applications of the Material/Energy Balance Principle. Wiley-Interscience, New York.

Baehr, H.D. 1965. Energie und Exergie. VDI-Verlag, Düsseldorf.

Berndt, E. 1985. From Technocracy to Net Energy Analysis: Engineers, Economists and Recurring Energy Theories of Value. In: Scott, A. (ed.). Progress in Natural Resource Economics. Clarendon Press, Oxford. pp 337-367.

Chapman, P.F. and Roberts, F. 1983. Metal Resources and Energy. Butterworths, London.

Cleveland, C.J. 1987. Biophysical Economics: Historical Perspective and Current Research Trends. *Ecological Modelling* 38:47-73.

Classen, R. S. and Girifalco, L. A. 1986. Materials for Energy Utilization. *Scientific American* 255:84-92.

Costanza, R. 1980. Embodied Energy and Economic Valuation. *Science* 210: 1219-1224.

Folke, C. and Kautsky, N. 1989. The Role of Ecosystems for a Sustainable Development of Aquaculture. *Ambio* 18:234-243.

Geogescu-Roegen, N. 1971. The Entropy Law and the Economic Process. Harvard University Press, Cambridge.

Gibbs, J.W. 1873. Collected Works. Yale University Press, 1948, New Haven. Originally published in *Trans. Conn. Acad.* 2:382-404.

Gilliland, M. 1978. Energy Analysis: A New Public Policy Tool. AAAS Selected Symposium 9. Westward Press, Boulder, Colorado.

Goldemberg, J., Johansson, T.B., Reddy, A.K.N. and Williams, R.H. 1988. Energy for a Sustainable World. John Wiley and Sons, New York.

Jansson, A.M. and Zucchetto, J. 1978. Energy, Economic and Ecological Relationships for Gotland, Sweden: A Regional Systems Study. *Ecological Bulletins* 28:154 pp.

Karlsson, S. 1982. The Exergy of Incoherent Electromagnetic Radiation. *Physica Scripta* 26:329-332.

Lindeman, R.L. 1942. The Trophic-Dynamic Aspects of Ecology. *Ecology* 23:399-418.

Lotka, A.J. 1924. Elements of Physical Biology. Williams and Wilkins, Baltimore, MD.

Martinez-Alier, J. 1987. Ecological Economics: Economics, Environment, Society. Basil Blackwell, Oxford.

Næss, A. 1973. Økologi, Samfunn og Livsstil. Universitetsforlaget, Oslo.

Odum, H.T. 1956. Primary Production in Flowing Waters. *Limnology and Oceanography* 1:102-117.

Odum, H.T. 1971. Environment, Power and Society. McGraw-Hill, New York.

Odum, H.T. 1972. An Energy Circuit Language for Ecological and Social Systems: Its Physical Basis. In: Patten, B.C. Systems Analysis and Simulation in Ecology, Vol II. Academic Press, London. pp 140-211.

Odum, H.T. 1984. Emergy in Ecosystems. In: Polunin, N. (ed.). Ecosystem Theory and Applications. John Wiley and Sons, New York. pp. 337-369.

Odum, H.T. 1988. Self-Organisation, Tranformity and Information. *Science* 242:1132-1139.

Odum, H.T. and Odum, E.C. 1976. Energy Basis for Man and Nature. McGraw Hill, New York.

Rawls, J. 1971. A Theory of Justice. Oxford University Press, Oxford.

Rant, Z. 1956. Exergi; ein Neues Wort für Technische Arbeitsfähigkeit. Forschung Ing.-Wesens 22:36-37.

Skolimowski, H. 1981. Eco-Philosophy: Designing a New Tactics for Living. Swedish Translation Eko-filosofi. Akademilitteratur, Stockholm 1982.

Slobodkin, L.B. 1960. Ecological Energy Relationships at the Population Level. *American Naturalist* XCIV:213-236.

Spreng, D.T. 1988. Net-Energy Analysis and the Energy Requirements of Energy Systems. Preager, New York.

Taylor, P.J. 1988. Technocratic Optimism, H.T. Odum and the Partial Transformation of Ecological Metaphor after World War II. *Journal of the History of Biology* 21:213-244.

Thomas, J.A.G. (ed.). 1977. Energy Analysis. IPC Science and Technology Press, Surrey.

Wall, G. 1986. Exergy: a useful concept. Ph.D. Dissertation in Physics. Physical Resource Theory Group, Chalmers University of Technology, Gothenburg.

Wall, G. 1987. Exergy Conversion in the Swedish Society. *Resources and Energy* 9:55-73.

Wall, G. 1990. Exergy Conversion in the Japanese Society. *Energy* 15:435-444.

Webb, M. and Pearce, D. 1975. The Economics of Energy Analysis. *Energy Policy* 3:318-331.

Linking the Natural Environment and the Economy; Essays from the Eco-Eco Group,
Carl Folke and Tomas Kåberger (editors)
Second Edition.
1992. Kluwer Academic Publishers

CHAPTER 5

Socio-Economic Dependence on the Life-Supporting Environment

by

Carl Folke

Beijer International Institute of Ecological Economics
The Royal Swedish Academy of Sciences
Box 50005, S-104 05 Stockholm, Sweden
and Department of Systems Ecology, Stockholm University

This chapter discusses the emerging ecology-economy perspective outlined by many systems ecologists and a growing number of economists. This perspective emphasizes the interrelations between socio-economic development and the life-supporting ecosystems. The role of the environment in supporting the economy is identified, and related to growth and sustainability issues. The challenge is to enable the agents of the human economy to fit the socio-economic systems into biogeochemical cycles, so as to maintain the life-supporting environment and hopefully extend the life of resources in fixed supply. Input management of human production systems and the development of ecotechnology are suggested as important tools for approaching such a goal.

1. INTRODUCTION

Nowadays, it has become evident to a growing number of people that socio-economic systems not only affect the environment but that they also depend on the life-supporting ecosystems in order to function. In a recent review of the book "Ecology and Our Endangered Life-Support Systems" (Odum, 1989a) it was stated that "despite all the advances in modern technology, society remains irrevocably dependent upon natural systems for life-support - a condition that is unlikely to change in the foreseeable future" (Ulanowicz, 1989).

The present chapter is about the significance of the life-supporting environment for socio-economic development. The first part defines the nature of the life-supporting environment. The next part relates this support to the emerging ecology-economy perspective of many systems ecologists and a growing number of economists. Finally, I discuss growth versus development and stress the need for input management of production systems and for the development of technologies that are in harmony with the life-supporting environment.

Throughout the paper it is emphasized that the "true" societal support value of ecosystems is connected to their physical, chemical, and biological role in the overall system, whether human preferences fully recognize that role or not (Clark & Munn, 1986; Costanza et al., 1989). It is argued that this economy-environment interrelation exists independently of ideological predisposition, but that it is affected differently depending on which ideological predisposition the policy is based upon.

2. THE LIFE-SUPPORTING ENVIRONMENT

There has been a tendency in society to focus more on life forms than on life processes. The value of nature has primarily been associated with the saving of endangered species, preferably large mammals, or with preservation of natural areas often of great recreational interest for sports fishing or hunting. Species perceived as having high value to humans are generally top consumers in the food-chains and parts of hierarchically self-organizing systems. For their survival they are dependent on life forms in lower food-chain levels, all the way back to the solar fixing plants and algae. Furthermore, organisms and their food-chains do not exist in isolation; they are dependent on and they modify the physical and chemical parts of their environment (Lovelock, 1979). The same is true for the human species, which for its survival is dependent on life forms in lower levels, while at the same time it modifies the life-supporting environment.

The life-support environment has been defined as that part of the earth that provides the physiological necessities of life, namely food and other energies, mineral nutrients, air, and water. The life-support ecosystem is the functional term for the environment, organisms, processes, and resources interacting to provide these physical necessities (Odum, 1989a).

From this perspective, one can divide ecosystems into three categories:

1. *Natural Environments* or Natural Solar-powered Ecosystems.

   These are the basic life-support systems, such as open oceans, upland forests, estuaries, wetlands, rain forests, lakes, rivers. They are self-supporting and self-maintaining, and some of them produce excess organic matter that may be exported to other systems or stored.

2. *Domesticated Environments* or Man-subsidized Solar-powered Ecosystems.

   These are the food and fiber producing systems such as agriculture lands, managed woodlands and forests, aquaculture. They are supported by industrial energy (e.g. fossil fuels, electricity), supplied by man, to run the tractors and to produce the fertilizers, fish pellets etc.

3. *Fabricated Environments* or Fuel-powered Urban-industrial Systems.

   These are mankind's wealth-generating, but also pollution-generating, systems, such as cities, suburbs, industrial areas, airports. In these systems the sun has been replaced as the major direct energy source by the use of fossil fuels and other industrial energies.

Using this division (Odum, 1975) it becomes evident that the *fabricated environments* are not self-supporting or self-maintaining. To be sustained they are dependent on the solar-powered *natural* and *domesticated environments* - the life-supporting ecosystems outside their own borders. Cities and their inhabitants, for example, have to receive energy, food, drinking water, and other natural resources from outside. For their daily survival they are dependent on large non-indigenous ecosystem areas (Borgström, 1967) for production of their food, for catching and purifying their drinking water, and so on. Such ecosystem-economy interrelations are especially obvious for densely populated countries such as Japan or Israel (Odum, 1989a). These countries could not be sustained without substantial imports of energy and matter from ecosystems outside their borders. Furthermore, it has been argued that ecosystem areas at least similar in size to those producing the inputs to an economy, a city, or a household, will, in due course, be required to process the outputs, that is the disposal of wastes and polluting substances from human activities (Folke et al., 1991).

2.1 Environmental Functions

Many of the environmental services[1] produced and sustained by the life-supporting environment are indispensable to humanity, such as the maintenance of the gaseous quality of the atmosphere and thus of the climate, operation of the hydrological cycle including flood control and drinking water supply, waste assimilation, recycling of nutrients, generation of soils, pollination of crops, provision of food from the sea, maintenance of a vast genetic library etc. (Ehrlich, 1989). de Groot (1988) has divided such environmental functions into four categories, namely: *Regulation functions*, *Production functions*, *Carrier functions*, and *Information functions* (see Table 1). However, it must be emphasized that none of these environmental functions can take place in isolation. *They are each the result of the dynamic and evolving structures and functions of their total ecological sub-system, and the fact that the socio-economic values of environmental functions and ecological sub-systems are directly connected to their physical, chemical, and biological role in the overall global system.*

Table 1. Life-support functions of the natural environment (modified from de Groot, 1988).

Regulation functions
- protection against harmful cosmic influences - climate regulation - watershed protection and water catchment - erosion prevention and soil protection - storage and recycling of industrial and human waste - storage and recycling of organic matter and mineral nutrients - maintenance of biological and genetic diversity - biological control - providing a migratory, nursery and feeding habitat
Production functions
- oxygen - food, drinking water, and nutrition - water for industry, households, etc. - clothing and fabrics - building, construction and manufacturing materials - energy and fuel - minerals - medical resources - biochemical resources - genetic resources - ornamental resources
Carrier functions
Providing space and a suitable substrate inter alia for: - habitation - agriculture,forestry, fishery, aquaculture

- industry
- engineering projects such as dams, roads
- recreation
- nature conservation

Information functions

- aesthetic information
- spiritual and religious information
- cultural and artistic inspiration
- educational and scientific information
- potential information

The very basis for all these functions is the solar-fixing ability of plants and algae. Without this solar-fixing ability the present self-organizing ecosystems, the human system included, would not be able to exist. This is one reason why ecologists sometimes use the fixed solar energy (gross primary production) as a measure of the life-support value of ecosystems (e.g. Gosselink et al., 1974; Costanza et al., 1989). Although the flow of solar energy is the essential driving force for life, self-organization and evolution, it is not sufficient by itself. "Matter matters too" (Georgescu-Roegen, 1971; Daly, 1984), as well as all the qualitative control mechanisms and feed-back loops within and between organisms and their environment (Holling, 1986). These intricate self-organizing and continuously evolving far from equilibrium systems are the result of the steady inflow of solar energy (Odum, 1983; Prigogine & Strengers, 1984).

2.2 Environmental Stress

Stress, caused by the disposal of wastes and pollutants, negatively affects recycling, feed-back loops, and control mechanisms in the life-supporting ecosystem, and thereby the production and maintenance of environmental goods and services. For example, a shortening of food chains with a reduction in top predators, such as salmonids, sea mammals, and birds of prey, frequently follows upon eutrophication and toxic waste pollution in fresh and marine waters. Such "disordering" disposals seem to alter species composition and increase the likelihood of overgrazing, parasitism, and other negative interactions in the ecosystem. Efficiency of resource use also decreases, and the internal cycling of nutrients and materials is reduced. The system becomes more open and leaky, more dependent on external inputs of energy and more difficult to predict, exploit and manage in a sustainable fashion (Odum, 1985). There exist several studies and modeling work aiming at a deeper understanding of the role of ecosystems and their response to perturbations in integrated ecology-economy systems (e.g. Odum, 1971; Rapport & Turner, 1977; Zucchetto & Jansson, 1985; Braat & van Lierop, 1987; Costanza et al., 1990).

3. THE ECONOMY AND THE LIFE-SUPPORTING ENVIRONMENT

3.1 The Conventional View

Especially since the Second World War the dominating approach has been to regard the economy as to superior and separated from whatever else is taking place in the environment. Land (natural resources and ecosystems included) has been viewed as one of three independent and substitutable primary factors of economic production.

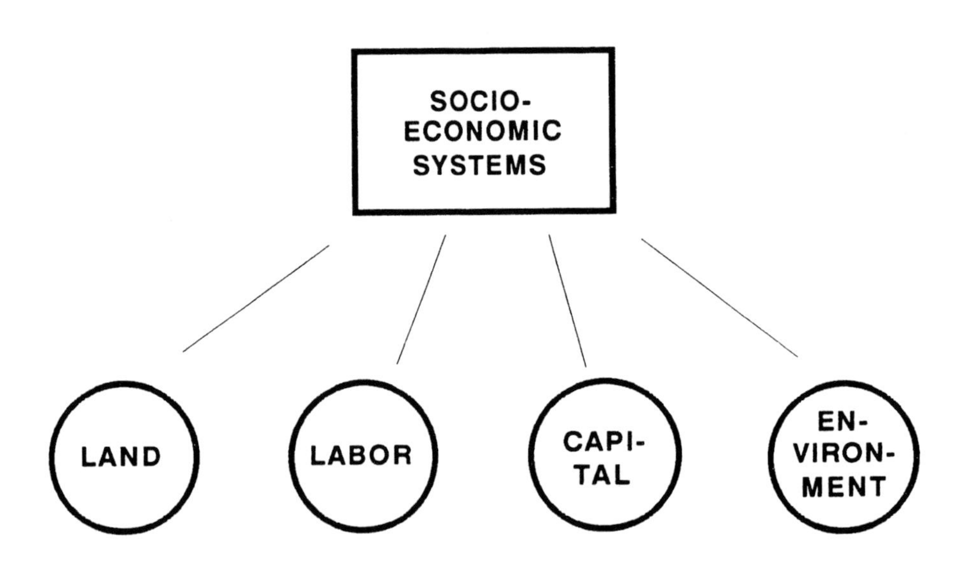

Figure 1. The dominating economy-environment perspective during industrial development. Land (natural resources), labor, and man-made capital have been viewed as the three independent primary factors of economic production. Within this perspective, the environment has been given higher priority in recent years

the other two being labor and capital (e.g. Barnett & Morse, 1963; Solow, 1974)(Figure 1). Land has generally been given less attention than the other two, and has been treated more with regard to its property relations (price of land, public ownership, community ownership, tenancy in city and country and so on) than as a complex physical input to economic production (Daly & Cobb, 1989). In recent years the environment (or more specifically environmental problems and concern) has achieved more attention, to a large extent due to the monetary costs associated with environmental degradation, and the economic benefits of pollution control and preservation of the environment. However, environmental issues have often been treated as occasional or incidental problems which need not cause concern until the environmental damage has occurred, or when there are obvious threats to human welfare, such as ozone depletion or acidification. Generally, environmental issues have been disconnected also from natural resource issues (Barbier, 1989), because the world outside the economy has been divided into two separate parts: *1)* Natural Resources and Energy, and *2)* Nature and the Environment, with management focusing either on the exploitation of natural resources, such as in agriculture, forestry or fishery, or on the preservation of endangered species, national parks and recreational options (Figure 2). This division (natural resources-environment) disconnects natural resource exploitation, and its positive and negative effects on ecosystems, from environmental quality issues necessary for the preservation of species. However, exploitation of natural resources and conservation of species are tightly linked, since resources, species, and their environments are integrated parts of ecosystems, they take place in the same life-supporting environment (Jansson, 1988).

It is still often argued among natural resource economists that resources are not, they become, because it is only when we learn to make use of, or expect to make use of, the components of the environment that they attain some economic value (e.g. Dasgupta, 1989). However, as discussed in the previous section of this chapter, we not only make daily use of unperceived and unacknowledged environmental goods and services, but we are for own existence also dependent on the life-supporting environment. Even if the environment has no price tag, it has an immense societal value. It sustains the economy by means of its life-support functions.

3.2 The Ecological Economic Perspective

Ecologists (e.g. Lotka, 1922; Odum, 1971; Odum, 1975; Zucchetto & Jansson, 1985; Holling, 1986; Ehrlich, 1989) and a growing number of economists (e.g. Boulding, 1966; Georgescu-Roegen, 1971; Daly, 1977; Hueting, 1980; Hufschmidt et al., 1983; Proops, 1985; Perrings, 1987; Barbier, 1989; James et al., 1989; Pearce & Turner, 1990) are well aware that the workings of ecosystems are pervasive in the economic system. "This pervasiveness arises from the simple fact that all economic activity uses up materials and resources and requires energy, and these, in turn, must end up somewhere - in dumps, dissipated in the atmosphere, disposed of to the oceans or whatever" (Pearce et al., 1989).

From an ecological-economic viewpoint, substitution between natural resources, and capital or labor is only possible in the short-term monetary perspective, and within a narrow system framework. In physical terms one cannot substitute either non-renewable or renewable natural resources with capital or labor. Capital and labor cannot physically create environmental goods and services. They are ultimately dependent on low-entropy energy and matter (Cleveland, 1987), since capital and labor themselves require energy, natural resources and services produced, maintained, and extracted from ecosystems. Hence, the monetary substitution is no substitution in physical terms.

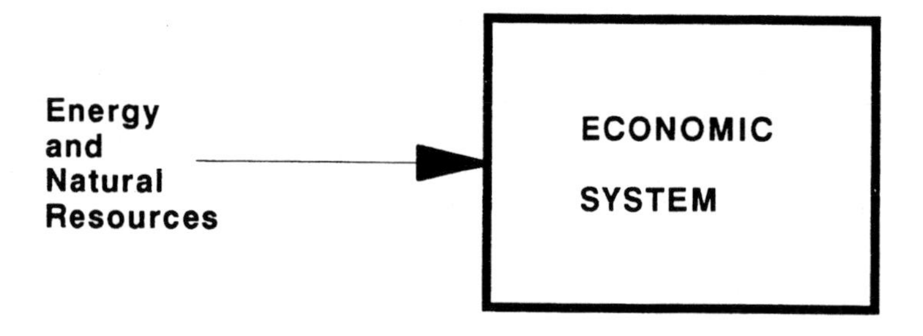

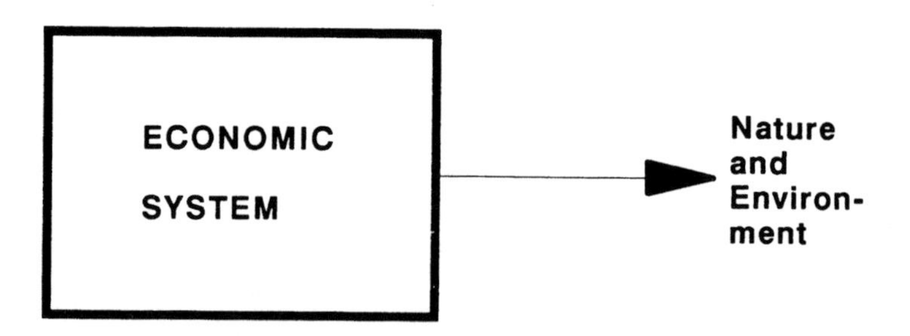

*Figure 2. The conventional division of the world outside the economy: **a)** natural resource management, **b)** environmental management. In reality **a)** and **b)** are tightly linked. They are dealing with integrated parts of the same life-supporting environment, and both alter the production and maintenance of ecosystem goods and services on which society depends (from Folke, 1990).*

Instead there is a *complementarity* between environmental goods and services and man-made capital, in the sense that they interact with each other and labor in the production of marketed goods and services in the economy (Costanza & Daly, 1990).

The perspective of substitution options makes it economically possible to replace goods and services from a damaged ecosystem with technical innovations. Although technical innovations may partly replace such goods and services, physically they are not independent substitutes, because to be produced and maintained they not only require energy and matter as such, but also the support from other ecosystems than the one that they are aimed at replacing.

The above discussion brings us to the ecology-economy perspective outlined in Figure 3. The economic system, like any ecological sub-system, is an open[2] dynamic system of the overall finite global ecosystem. The two are physically connected by the throughput of energy and matter from ecosystem sources and by other environmental goods and services sustaining economic activity (see section 2). Solar energy drives the production of all ecosystem goods and services, and industrial energy (fossil fuels, electricity etc.) is used through the processes within the industrial economy to "produce" (upgrade) matter to consumable commodities. Energy and matter are eventually returned to ecosystems as heat and waste. This means that economic production of any commodity needs natural resources, and the transformation of natural resources, from discovery, extraction, refinement and so on, into useful raw materials and eventually into humanely produced goods and services, requires the use of industrial energy (Hall et al., 1986) as well as the support by ecosystems driven by solar energy (Odum, 1989a).

From this ecological economic perspective, the expansion of the economic sub-system is limited by the size of the overall finite global ecosystem, by its dependence on the life-support sustained by intricate ecological connections which are more easily disrupted as the scale of the economic sub-system grows relative to the overall system (Odum, 1971; Daly, 1984, 1987a, b).

Hence, the economic sub-system rests on its physical foundations, such as the laws of thermodynamics, and thus must behave, to be sustained, in a way that is consistent with these physical laws (Georgescu-Roegen, 1971; Odum, 1971). In contrast to Figure 2, it becomes evident that exploitation of natural resources and disposal of wastes takes place in the same environment, and that both modify the contributions and support to society provided by ecosystems. It is also important to realize that the interrelations between ecosystems and economic systems concern the dynamic structure, function, and performance of compartments in *both* systems and the flows and feed-backs between all these compartments. These compartments include also significant institutional, political, cultural and social factors through which action is carried out.

3.3 The Scale Problem

During the industrial development it was easy to take the life-support from ecosystems for granted, because human activities were on too small a scale relative to environmental processes and functions to interfere with the free provision of ecosystem goods and services, the waste assimilative capacity included. As stated by Perrings (1987), "In the short period, the economic system, like any other sub-system of the global system, is not bound by the law of conservation of mass. That is, the physical mass of economically valued resources can expand at the expense of the environment. Eventually, however, the economy will run up against the constraints to growth imposed by the law of conservation of mass." Hence, it is impossible for any sub-system, the human economy included, to expand indefinitely in a finite world (Odum, 1973; Daly & Cobb, 1989), especially since economic activities significantly can reduce the capacity of the life-supporting environment to yield the flow of necessary environmental goods and services.

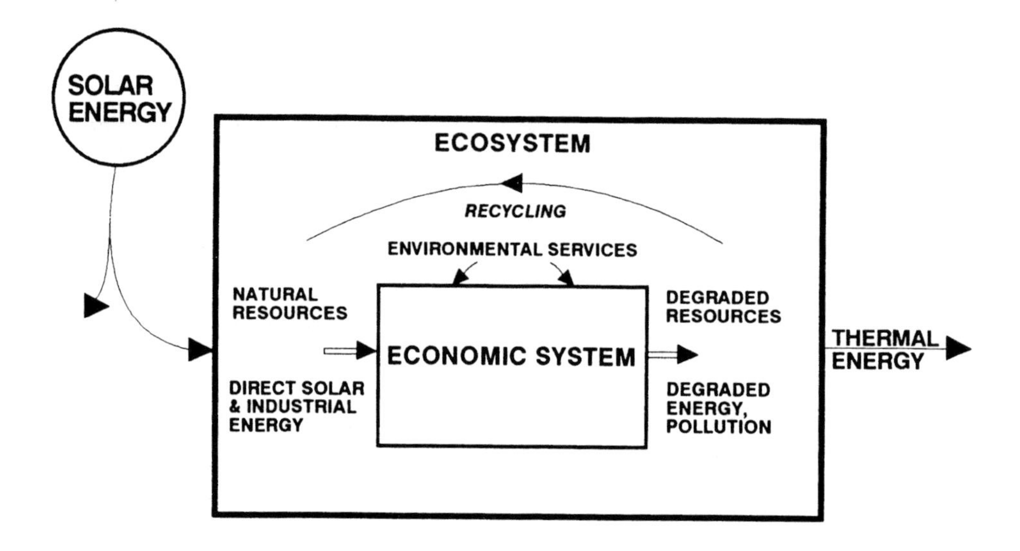

Figure 3. The economic system is an open sub-system of the overall ecosystem. To make industrial development and technological change possible, ecosystems are used as sources for energy and natural resources and as sinks for waste outputs, and these activities affect the ecosystem contributions of environmental goods and services. As the scale of socio-economic systems have expanded relative to the ecosystems, economy-environmental interdependence and complementarity have become more obvious (from Folke, 1990).

Since the human expansion, with the associated exploitation and disposal of wastes and pollutants, not only affects the natural environment as such, but also the level and composition of environmentally produced goods and services required to sustain society, the economic sub-system will be limited by the impacts of *its own actions* on the environment. If this interdependence is not recognized the limit for economic activity will become lower and the possibilities for sustainable development and economic growth more constrained. However, there are potentials for a more sustainable allocation of resources within these limits, and this is further discussed in relation to the development and growth concepts at the end of this chapter. This perception of the life-supporting environment is essential if a more efficient resource allocation is to be achieved.

To sum up, environmental life-support and technological change are complementary rather than substitutable and socio-economic development is dependent on both. Today, environmental goods and services (the natural capital) are increasingly replacing man-made capital as the limiting factor for economic development (Daly & Cobb, 1989; Daly, 1990). It is only possible to approach a sustainable development with this broadened ecological-economic framework and a longer time-scale perspective. Economic analysis cannot be abstracted from the wider physical environment in which production takes place (Perrings, 1987).

4. CONVENTIONAL EXPLOITATION VERSUS SUSTAINABLE USE OF THE LIFE-SUPPORTING ENVIRONMENT BY THE USE OF INPUT MANAGEMENT AND ECOTECHNOLOGY

4.1 Resource Exploitation and Substitution

It is possible to substitute some natural resources for other natural resources, since it is the properties and functions of a resource that are generally required in economic production, not the uniqueness of a particular resource. This is one reason for the exhaustion or extinction of natural resources and the destruction of ecosystems. However, because of the uniqueness of matter, substitution of any one resource normally requires replacement of the technology by which resources dependent on its use are produced and in the character of those resources (Perrings, 1987). This has been illustrated for the exploitation of several resources (Berkes, 1989). One example is from the New Brunswick forests, Canada, where the progressive exploitation started with the most accessible high quality trees. When they were exhausted it continued with the next highest quality wood and so on, and thus continuously degraded the quality of natural resources provided by the forest ecosystem. The resource-based industries developed, but the forest ecosystem had deteriorated, because the local and regional economies were managed as if they were independent on their environment (Reiger & Baskerville, 1986). The degradation required innovations and development of more and more capital, resources, and fossil fuel intensive technologies for the production of quality products out of raw materials of lower and lower quality. A similar pattern has been described for the exploitation of various resources in the U.S. (Cleveland, 1990). Another example is from the Pacific northwest where the salmon fisheries followed a predictable sequence of geographical expansion from more accessible to less accessible areas, and from the more valuable to the less valuable species (Rogers, 1979). This general pattern of *sequential exploitation* (from the more accessible to increasingly less accessible areas and from the most valuable to the less and less valuable species or lower and lower quality resources) appears to be a common feature of resources development in general (Grima & Berkes, 1989), and when resources and ecosystems have become degraded often results in severe social, political and economic conflicts. In New Brunswick, it has finally been realized that the prosperous growth of both the local and regional economy would not have been possible without the life-support provided by the forest ecosystem. The progressive exploitation of the ecosystem to expand the economy has today made it necessary to redevelop the ecosystem in order to maintain the economy (Reiger & Baskerville, 1986). The design of the solution to such problems must proceed from the regional perspective based on an integrated dynamic ecological-economic perspective.

A similar pattern of exploitive development is presently repeated in the rapidly expanding fish farming industry. The focus is on increased production at the local rearing site, sectoral management and physical growth predictions. The capacity of the ecosystem to support the expanding industry with resources and to process its waste is seldom accounted for (Folke & Kautsky, 1989).

4.2 Growth and Development

As illustrated by the examples above, there certainly exists a point in growth of the economy beyond which sustainability is no longer possible because of ecological limits. This brings us into the largely debated growth issue. Once again Daly (1987b) has contributed by defining the term growth. "By *growth* I mean quantitative increase in the scale of the physical dimension of the economy; i.e. the rate of flow of matter and energy through the economy, and the stock of human bodies and artifacts. By *development* I mean the qualitative improvement in the structure, design, and composition of physical stocks and flows, that result from greater knowledge, both of technique and of purpose." Although there are different views concerning the potentials for a more efficient use of industrial energy (e.g. Cleveland et al., 1984; Goldemberg et al., 1988), the present use of natural resources is highly inefficient, and perhaps more important, there is a large potential for recycling of resources and reduction of waste and pollutants (Figure 3). This means that there is a potential for economic growth based on development (qualitative improvement) that includes the perception of the socio-economic necessity of the life-supporting environment. Such efficiency-increasing development is very different from the throughput-increasing growth we have seen during the recent century.

Severe economic and societal costs are the result when local and regional environmental limits are exceeded, as illustrated by the pervasive environmental degradation in many countries. That we are approaching environmental limits on a global scale is reflected in warning signals such as global warming and ozone depletion. The closer we are to limits the less we can assume that economic welfare, as generally defined today, and total welfare move in the same direction. Rather we must learn to define and explicitly account for the other sources of welfare that growth inhibits and erodes when it presses against limits (Daly, 1987b). As has been shown in this chapter the life-supporting environment is such an essential source of welfare.

4.3 Input Management and Ecotechnology

From a systems ecology perspective, two important keys have recently been suggested for integrating the economy with the life-supporting environment. One is *Input Management* (Odum, 1989b), the other is *Ecotechnological Development* (Mitsch & Jörgensen, 1989).

Input Management shifts the emphasis to the input side of production systems. Costly and environmentally damaging inputs can be reduced with concurrent reductions in non-point pollution, by increasing the efficiency of resource use, minimizing the use of polluting materials and chemicals, and recycling, without major sacrifices in yields of food, electricity, or manufactured goods (Odum, 1989b).

Ecotechnology, or Ecological Engineering, implies the design of human society with its natural environment for the benefit of both, and it is thus the opposite of many other forms of engineering and technology that try to substitute or conquer the natural environment. Ecotechnology considers an assemblage of species and their abiotic environment as a self-organizing system that can adapt to changes brought about by outside forces (Mitsch & Jörgensen, 1989). The idea is to use small amounts of supplementary industrial energy to control ecosystems in which the main energy drives are still coming from natural sources (Odum, 1971).

At present, a large share of our use of industrial energy and other natural resources is required to replace environmental goods and services which have deteriorated (e.g. Folke, Chapter 8 this book). Recognizing the societal value of ecosystems, and making use of them by supplementing and enhancing their production and maintenance of environmental goods and services, not only makes the use of costly and often polluting industrial energy more efficient and redirects it to other parts of the economy, but also leads to conservation of species and ecosystems. Ecotechnology, also makes it possible, through recycling, to reinvest part of the non-renewable resources, used up in the economy into environmental goods and services, instead of having them accumulating in dumping sites or in land fills.

4. CONCLUSION

It is sometimes argued that economic development is sustainable if the economic benefits of present human actions are larger than the monetary costs for environmental degradation. In the ecological economic perspective development is sustainable when the economic systems are adjusted within the framework of the overall finite global system. Economic systems require energy and matter and the life-supporting environment for maintenance and development. Economies cannot exist by themselves.

If management and policies emphasize optimal resource allocation in the economy only, without considering the effects of exploitations or the disposal of wastes and pollutants, ecosystems and their life-support will be continuously degraded. The world economy and human population is approaching magnitudes at which its present activities can no longer be absorbed by the life-support systems without better integral couplings of its material cycles (Odum, 1973). Perceptions of the socio-economic significance of the life-supporting environment are evolving (e.g. WCED, 1987). The challenge is to perceive its significant role and enable the agents of the human economy to fit the socio-economic systems into biogeochemical cycles, so as to maintain and hopefully to extend the life of resources in fixed supply. Perception of the "real" value of the life-supporting environment, and maintenance of its capability to sustain human systems are the true basis for sustainability.

ACKNOWLEDGEMENTS

Many thanks especially to Ann Mari Jansson for her support, enthusiasm and encouragement through the years we have worked together, for letting me share her extensive knowledge on complicated matters in the interface of ecology and economics, and for creating a scientifically very stimulating environment from which the perspective presented here is a part. I am also grateful to the referees Herman Daly, Frank Golley and Rudolf de Groot, and also to Folke Günther and Monica Hammer who provided constructive comments on the manuscript. This study was funded by the Swedish Council for Forestry and Agricultural Research (SJFR), and the the Swedish Council for Planning and Coordination of Research (FRN), which help is gratefully acknowledged.

NOTES

1. *Environmental goods* refer to both non-renewable, such as oil, ore, and renewable natural resources (ecosystem based) such as fish, wood, drinking water and so on. *Environmental services* refer to maintenance of the gaseous quality of the atmos-

phere, of climate, operation of the hydrological cycle including flood controls and drinking water supply, waste assimilation, recycling of nutrients, generation of soils, pollination of crops, provision of food from the sea, maintenance of a vast genetic library and so on. Non-renewable environmental goods are extracted from the life-supporting environment, and renewable environmental goods and environmental services are produced and maintained in the life-supporting ecosystems.

2. The global system is a closed system since it only exchanges energy with its surroundings. All subsystems within the global system, such as the human economy, are open systems with respect to their environments. All subsystems exchange both matter and energy with their environments.

REFERENCES

Barbier, E.B. 1989. Economics, Natural-Resource Scarcity and Development: Conventional and Alternative Views. Earthscan, London.

Barnett, H.J. and Morse, C. 1963. Scarcity and Growth. Johns Hopkins University Press, Baltimore.

Berkes, F. 1989. Common Property Resources: Ecology and Community-based Sustainable Development. Belhaven Press, London.

Borgström, G. 1967. The Hungry Planet. Macmillan, New York.

Boulding, K.E. 1966. The Economics of the Coming Spaceship Earth. In: Daly, H.E. (ed.), 1980. Economics, Ecology, Ethics: Essays Toward a Steady-State Economy. Freeman, San Francisco. pp. 253-263.

Braat, L.C. and van Lierop, W.F.J. 1987. Economic-Ecological Modeling. North-Holland/Elsevier, Amsterdam.

Clark, W.C. and Munn. R.E. 1986. Sustainable Development of the Biosphere. Cambridge University Press, Cambridge.

Cleveland, C.J. 1987. Biophysical Economics: Historical Perspective and Current Research Trends. *Ecological Modelling* 38:47-73.

Cleveland, C.J. 1990. Natural Resource Scarcity and Economic Growth Revisited: Economic and Biophysical Perspectives. Presented at the International Society for Ecological Economics' Symposium on The Ecological Economics of Sustainability: Making Local and Short-term Goals Consistent with Global and Long-term Goals. The World Bank, Waskington, May 1990.

Cleveland, C.J., Costanza, R., Hall, C.A.S. and Kaufmann, R. 1984. Energy and the US Economy: A Biophysical Perspective. *Science* 225:890-897.

Costanza, R. and Daly, H.E. 1990. Natural Capital and Sustainable Development. Workshop on Natural Capital, March 15-16, 1990. Canadian Environmental Assessment Research Council, Vancouver, Canada (mimeographed).

Costanza, R., Faber, S.C. and Maxwell, J. 1989. Valuation and Management of Wetland Ecosystems. *Ecological Economics* 1:335-361.

Costanza, R., Sklar, F.H. and White, M.L. 1990. Modeling Coastal Landscape Dynamics. *BioScience* 40:91-107.

Daly, H.E. 1977. Steady-State Economics. Freeman, San Francisco.

Daly, H.E. 1984. Alternative Strategies for Integrating Economics and Ecology. In: Jansson, A.M. (ed.). Integration of Economy and Ecology: An Outlook for the Eighties. Proceedings from the Wallenberg Symposia. Askö Laboratory, Stockholm University, Stockholm. pp. 19-29.

Daly, H.E. 1987a. Filters Against Folly in Environmental Economics: The Impossible, the Undesirable, and the Uneconomic. In: Pillet, G. and Murota, T. (eds.). Environmental Economics: The Analysis of a Major Interface. R. Leimgruber, Geneva. pp. 1-10.

Daly, H.E. 1987b. The Economic Growth Debate: What some Economists have learned but Many have not. *Journal of Environmental Economics and Management* 14:323-336.

Daly, H.E. 1990. Toward Some Operational Principles of Sustainable Development. *Ecological Economics* 2:1-6.

Daly, H.E. and Cobb, J.B. 1989. For the Common Good: Redirecting the Economy Toward Community, the Environment and a Sustainable Future. Beacon Press, Boston.

Dasgupta, P. 1989. Exhaustible Resources. In: Friday, L. and Laskey, R. (eds.). The Fragile Environment. Cambridge University Press, Cambridge. pp. 107-126.

de Groot, R.S. 1988. Environmental Functions: An Analytical Framework for Integrating Environmental and Economic Assessment. Workshop on Integrating Environmental and Economic Assessment: Analytical and Negotiating Approaches, November 17-18, 1988. Canadian Environmental Assessment Research Council, Vancoucer, Canada (mimeographed).

Ehrlich, P.R. 1989. The Limits to Substitution: Meta-Resource Depletion and a New Economic-Ecological Paradigm. *Ecological Economics* 1:9-16.

Folke, C. 1990. Evaluation of Ecosystem Life-Support: In Relation to Salmon and Wetland Exploitation. Ph.D. Dissertation in Ecological Economics. Department of Systems Ecology, Stockholm University, Stockholm.

Folke, C. and Kautsky, N. 1989. The Role of Ecosystems for a Sustainable Development of Aquaculture. *Ambio* 18:234-243.

Folke, C., Hammer, M. and Jansson, A.M. 1991. The Life-Support Value of Ecosystems: A Case Study of the Baltic Sea Region. *Ecological Economics* 3: in press.

Georgescu-Roegen, N. 1971. The Entropy Law and the Economic Process. Harvard University Press, Cambridge.

Goldemberg, J., Johansson, T.B., Reddy, A.K.N. and Williams, R.H. 1988. Energy for a Sustainable World. John Wiley and Sons, New York.

Gosselink, J.G., Odum, E.P. and Pope, R.M. 1974. The Value of a Tidal Marsh. Publication No. LSU-SG-74-03. Center for Wetland Resources, Louisiana State University, Baton Rouge.

Grima, A.P.L. and Berkes, F. 1989. Natural Resources: Access, Right-to-use and Management. In: Berkes, F.(ed.). Common Property Resources: Ecology and Community-based Sustainable Development. Belhaven Press, London. pp. 33-54.

Hall, C.A.S., Cleveland, C.J. and Kaufmann, R. 1986. Energy and Resource Quality: The Ecology of the Economic Process. John Wiley and Sons, New York.

Holling, C. 1986. The Resilience of Terrestrial Ecosystems: Local Surprise and Global Change. In: Clark, W.C. and Munn, R.E. (eds.). Sustainable Development of the Biosphere. Cambridge University Press, London. pp. 292-317.

Hueting, R. 1980. New Scarcity and Economic Growth: More Welfare through less production. North-Holland, Amsterdam.

Hufschmidt, M.M., James, D.E., Meister, A.D., Bower, B.T. and Dixon, J.A. 1983. Environment, Natural Systems, and Development: An Economic Valuation Guide. Johns Hopkins University Press, Baltimore.

James, D.E., Nijkamp, P. and Opschoor, J.B. 1989. Ecological Sustainability and Economic Development. In: Archibugi, F. and Nijkamp, P. (eds.). Economy and Ecology: Towards Sustainable Development. Kluwer Academic Publishers, Dordrecht. pp. 27-48.

Jansson, A.M. 1988. The Ecological Economics of Sustainable Development: Environmental Conservation Reconsidered. In: The Stockholm Group for Studies on Natural Resources Management (eds.). Perspectives of Sustainable Development: Some Critical Issues Related to the Brundtland Report. Stockholm Studies in Natural Resources Management No. 1. Askö Laboratory, Division of Natural Resources Management, Stockholm University, Stockholm. pp. 31-36.

Lotka, A.J. 1922. Contributions to the Energetics of Evolution. *Proceedings from the National Academy of Science* 8:147-151.

Lovelock, J.E. 1979. Gaia: A New Look at Life on Earth. Oxford University Press, Oxford.

Mitsch, W.J. and Jörgensen, S.E. 1989. Ecological Engineering: An Introduction to Ecotechnology. John Wiley and Sons, New York.

Odum, E.P. 1975. Ecology: The Link Between the Natural and Social Sciences. Second Edition. Holt-Saunders. New York.

Odum, E.P. 1985. Trends to be Expected in Stressed Ecosystems. *BioScience* 35:419-422.

Odum, E.P. 1989a. Ecology and Our Endangered Life-Support Systems. Sinuaer Associates, Sunderland, Massachusetts.

Odum, E.P. 1989b. Input Management of Production Systems. *Science* 243:177-181.

Odum, H.T. 1971. Environment, Power, and Society. John Wiley and Sons, New York.

Odum, H.T. 1973. Energy, Ecology, and Economics. *Ambio* 2:220-227.

Odum, H.T. 1983. Systems Ecology. John Wiley and Sons, New York.

Pearce, D.W. and Turner, R.K. 1990. Economics of Natural Resources and the Environment. Harvester-Wheatsheaf, London.

Pearce, D., Markandya, A. and Barbier, E.B. 1989. Blueprint for a Green Economy. Earthscan, London.

Perrings, C. 1987. Economy and Environment. Cambridge University Press, Cambridge.

Prigogine, I. and Strengers. I. 1984. Order out of Chaos. Bantam Books, New York.

Proops, J.L.R. 1985. Thermodynamics and Economics: From analogy to Physical Functioning. In: van Gool, W. and Bruggink, J.J.C. (eds.). Energy and Time in the Economic and Physical Sciences. North-Holland, Amsterdam. pp. 155-174.

Rapport, D.J. and Turner, J.E. 1977. Economic Models in Ecology. *Science* 195:367-373.

Reiger, H.A. and Baskerville, G.L. 1986. Sustainable Redevelopment of Regional Ecosystems Degraded by Exploitive Development. In: Clark, W.C. and Munn, R.E. (eds.). Sustainable Development of the Biosphere. Cambridge University Press, London. pp. 75-101.

Rogers, G.W. 1979. Alaska's Limited Entry Program: Another View. *Journal of the Fisheries Board of Canada* 36:738-788.

Solow, R.M. 1974. The Economics of Resources or the Resources of Economics. *The American Economic Review* 64:1-14.

Ulanowicz, R.E. 1989. Book Review of Ecology and Our Endangered Life-Support Systems. *Ecological Economics* 1:363-365.

Zucchetto, J. and Jansson, A.M. 1985. Resources and Society: A Systems Ecology Study of the Island of Gotland, Sweden. Springer Verlag, Heidelberg.

WCED. World Commission for Environment and Development. 1987. Our Common Future. Oxford University Press, Oxford.

Part II

THE ROLE AND VALUE OF THE NATURAL ENVIRONMENT

Linking the Natural Environment and the Economy;
Essays from the Eco-Eco Group,
Carl Folke and Tomas Kåberger (editors)
Second Edition.
1992. Kluwer Academic Publishers

CHAPTER 6

Ecological Consequences of Long-Term Landscape Transformations in Relation to Energy Use and Economic Development

by

Ann Mari Jansson

Department of Systems Ecology
Stockholm University
S-106 91 Stockholm, Sweden

Changes in the overall productivity of terrestrial ecosystems were evaluated in a historical analysis of landscape transformations in the island of Gotland, Sweden. By comparing the total amount of solar energy fixed in photosynthesis in the 1980's and in pre-industrial times it was found that the gross primary production had decreased by 3.7 per cent. This was primarily due to the extensive drainage of mires and the reclamation of deciduous forests related to a five-fold expansion of the agricultural area. With the introduction of modern farming machinery, artificial fertilizers and irrigation techniques, the yearly net yield in agriculture increased from 1 to 3.4 tonnes per hectare. In 1972, the subsidy of fossil energies to agriculture was 30 times larger than the solar energy contributions, expressed in units of similar quality. It is predicted that the negative impacts on water resources and the depletion of peat layers due to intensified agriculture will constrain the development of the island's economic system as fuels and other purchased resources become more scarce in the future. The historical analysis reveals a pulsing pattern in the development of human activities on the island. It is concluded that these pulses to a large extent are related to the availability of external resources as well as cultural influences. The results are suggested to reflect a general pattern of prosperous growth and decay in Man-Nature relations.

1. INTRODUCTION

During recent decades, many ecologists have become involved in studying how human activities have transformed ecosystems, landscapes, geographical regions and the whole biosphere. The linked problems of economic development and ecological stability, which were brought to general attention by the Brundtland report in 1987 (WCED, 1987), placed new demands on ecologists to help to explain the potentials and ramifications of the natural environment in human affairs. Many members of the public are now looking to ecologists to provide policy guidance on how to make economic production harmonize with natural processes. Strategies for using the products and services of the biosphere are important considerations in environmental management.

To facilitate evaluation and prediction of the work that nature performs, analysis of the flows of energy and materials between ecological and economic systems has emerged as a necessary complement to economic accounting (e.g. Cottrell, 1955; Odum, 1971; Odum & Odum, 1976; Ayres, 1978; for reviews see Cleveland, 1987; Martinez-Alier, 1987). Knowledge of the relationship between energy and the economic functioning is also important in order to be able to improve planning with regard to uncertainties in the supply and price of non-renewable resources, particularly fossil energy (e.g. Pearson, 1989).

During the last two hundred years the use of fossil energies and electricity in the industrial society has replaced much of the free solar energy powered work of natural and semi-natural ecosystems. Stimulated by large inputs of auxiliary energies in the form of fertilizers, pesticides, irrigation water and fuel-powered machinery an increasing part of the terrestrial production of organic material has been channelled into food and timber products. Vitousek et al. (1986) estimated that, on a global scale, humans now harvest or appropriate about 40 per cent of the products of photosynthesis on land. Such human domination of biomass production restricts the capacity to survive for other species, many of which are now threatened by extinction, and which contribute in maintaining the life-supporting environment on which humans depend. This also leads to the depletion of natural capital stocks such as fertile soils and clean water and to the accumulation of wastes which cannot be absorbed by the environment.

Taken together, the human-induced changes in terrestrial ecosystems play a large part in the process leading to increased concentrations of carbon dioxide, methane and nitrogen oxides in the atmosphere and thereby on the control of the global energy balance (Bolin, 1986, 1989). The socio-economic and political fluctuations that would result from possible climatic changes or constraints on the use of fossil energies are difficult to predict, but this uncertainty makes it urgent to examine, for a variety of regions, how the solar energy based terrestrial production can be restored in order to strengthen the buffering capacity of society's life-supporting ecosystems.

The aim of this chapter is to explore - using the island of Gotland in the Baltic Sea as a specific example - how the potential work contribution of terrestrial ecosystems to a regional economy has changed with time as a result of human land use. Comparisons are made of the potential useful work of ecosystems, as indicated by estimates of the gross primary production of the regional mosaic of ecosystems in present and pre-industrial times. The main purpose is to use some of the results of a rather extensive research project on Gotland's ecological and economic systems (Jansson & Zucchetto, 1978a,b; Zucchetto & Jansson, 1979, 1985) to generate insight on how to manage land and water resources for an ecologically sustainable economic development.

2. THE ENERGY SPECTRUM AND THE NOTION OF CARRYING CAPACITY

Ecologists have often chosen islands for investigating ecological principles and the causal relationships between natural and man-made systems. The advantages are obvious. There are well defined boundaries and it is relatively easy to keep track of the inflows and the outflows. Islands often exhibit a diversity of ecosystems and human activities that make them convenient model systems for testing ideas on such complex issues as the carrying capacity of the biophysical resources with regard to human activities (Jansson, 1985). For a subsistence economy this is a relatively easy task since the carrying capacity depends on the potential productivity of the existing ecosystems, but the emergence of trade allows the human settlement to be supported by a vastly larger area than the given territory. Nevertheless, there are inherent limits of the capacity on the indigenous resources to supply clean air and water and maintain a diverse, natural environment for a growing human society. Such regional limits might be easier to identify for an island, that can provide insights about the interdependence between the ecological and the economic systems.

In one respect islands are different from mainland regions, however. The length of the coastline makes the contributions from wind and wave energy comparatively larger than for most mainland regions. The energy spectrum for the island of Gotland (Table 1), expressed in solar-energy equivalents, shows that the wave energy accounts for almost 75 per cent of the total renewable energy while the energy associated with freshwater makes up only a small fraction. A superficial assessment of the present composition of Gotland's economic system would not emphasize that the high physical energy-flows of the coastal waters constribute to a substantial degree to the economic growth. However, there are many less obvious benefits of wave energy. The sewage outlets and polluting effluents in rivers and land runoff are dispersed by the action of waves at no cost to society. Waves help to maintain and clean the sandy beaches and rocky shores which are important attractions for the many tourists who visit Gotland in the summer and who add considerably to the economy.

3. THE EVOLUTION OF A LANDSCAPE PATTERN

Gotland is a limestone island of some 3,100 km^2 located in the Baltic Sea. The terrestrial systems on Gotland have developed since the glaciers of the last ice-age receded from the Baltic coast some 10,000 years ago resulting in a continuous elev-

Table 1. Natural and industrial energy flows for Gotland, 1972 (PJ, in solar equivalents) (from Zucchetto & Jansson, 1985).

Type of energy		PJ
Natural ("free") energies		
	Wave energy	514,080
	Gross photosynthesis	92,000
	Wind energy	47,880
	Salt/freshwater	41,400
	Head of water	2,040
Total		**697,400**
Industrial (purchased) energies		**141,606**

ation of the land and subsequent fluctuations in climate. The initial tundra ecosystem with large flooded and swampy areas was succeeded by birch, asp and pine forests. With the gradual warming, the length of the growing season increased and trees such as oaks, hazels, elms and lindens invaded. Within a relatively short time span, the newly exposed land surface had self-organized and evolved into a complex system of mixed deciduous and coniferous forests, wetlands in shallow basins of the bedrock, and a sparse heath vegetation on flat limestone areas with only a thin mantle of soil.

Figure 1 illustrates the development of human activities on Gotland, showing periods of prosperous growth and decay. The first human immigration to the island, of seal-hunters and fishermen, took place during the late Stone Age about 7,000 years ago. A slash-and-burn type agriculture emerged in the early Bronze Age (2,000-1,000 BC) primarily in the deciduous forests, where energy stored in biomass and organic soils could be exploited. The location of these early human settlements can still be identified by means of high contents of phosphate in the soils (Arrhenius, 1955).

After a climate decline during the early Iron Age human residence and cultivation became more stationary. Domestic animals and the deciduous pastures were the main basis for survival of the farm household. The remains of some 2,000 stone-house foundations, fencing systems and pre-historic fields in Gotland provide information about the density of human settlements and cultivation patterns during the Iron Age (AD 200-600) indicating that about 100 km^2 of pasture land and 27 km^2 of crop land supported a resident population of 3,000-5,000 people in pre-historical times in Gotland (Carlsson, 1979).

In the fifth century, the time when the Roman Empire collapsed, Gotland experienced a period of recession but by the beginning of the Viking period (AD 800), the agrarian system had recovered and was able to produce a surplus for export overseas. Numerous silver coins from various European countries found in Gotland's soils give evidence of this early trade. By the end of the first millennium the size of the resident population was about 10,000.

In the 13th century, German tradesmen of the famous Hanseatic League in the Baltic region made the town of Visby on Gotland a wealthy center of trade. The island had about 13,000 inhabitants at that time, more than half living in Visby (Jansson, 1985). A divergence of wealth and the antagonism between the island's urban and rural populations together with outbreaks of disease and war put an end to this period of prosperity in Gotland's history. For three hundred years the island was almost totally isolated and constrained by its local resources. Many farms were abandoned (Ersson, 1974) indicating a serious decrease in the farming population.

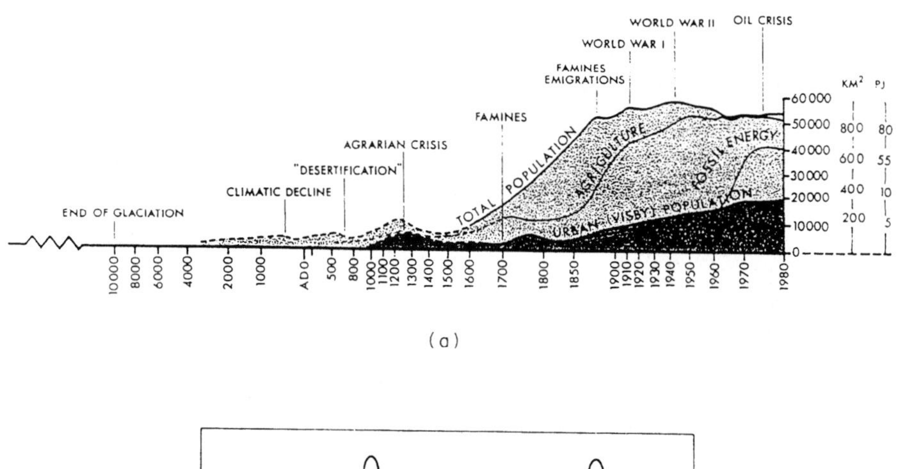

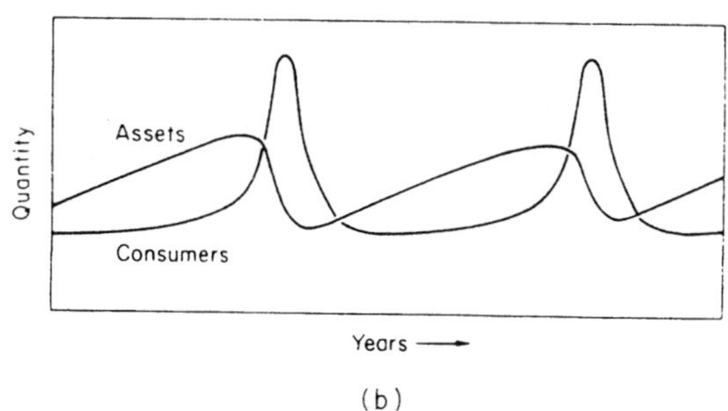

Figure 1. a) Human development in Gotland from pre-history to 1980. The figure illustrates a pulsing pattern in the development of human activities. Solid curves illustrate actual development, dotted curves are reasonable estimates based on historic documents, archaeologic evidence, and climate change (modified after Zucchetto & Jansson, 1985). b) A pulsing pattern suggested as a general phenomena in the development of ecosystems as well as in man-nature relationships (Odum, 1982, 1988).

4. RECLAMATION OF LAND AND POPULATION GROWTH

A new period of growth and development began in the seventeenth century, after the introduction of new technologies in the form of water-wheels, water-saws and windmills allowed the transformation of energy from water-flow and wind into mechanical work. The early industrialization was based on local resources of limestone and forestry products produced for export. This exploitation had a negative impact on the storage of organic matter in the terrestrial ecosystems. Open spaces in areas of forested heath still bear witness to the intensive harvesting of firewood and the occasional forest fires associated with lime-burning and the production of tar in Gotland several hundred years ago.

With the introduction of iron ploughs and new types of harrows former grazing lands could be used for the production of grain. This in turn forced the conversion of forests into pastures for the high number of cattle needed to furnish enough manure and draught animals to support agricultural production on arable lands. The rapid process of land reclamation led to dynamic economic and social changes. Between 1600 and 1700 the human population in Gotland almost doubled in number (Figure 1) without any clear evidence of actual poverty or scarcity of food.

Valuable information about land use and agricultural production around the year 1700 can be extracted from the map drawn by Swedish authorities as a basis for taxation. The total area of arable land was 195 km^2 in 1700, but between one-third and one-half of that area seems to have lain fallow since a 2-3 year crop rotation system was practiced (Moberg, 1938). Animal production was the most important part of the peasant economy and the number of cattle and draught-animals was as high as 150 animals per km^2 of cultivated land (Table 2).

By the beginning of the nineteenth century the human population had increased to 31,000 of whom close to 90 per cent still resided in the countryside. The nineteenth century was characterized by large-scale reclamations of land. The productivity of the island's extensive wetland ecosystems and the organic matter stored in their peat soils now became appropriated for agricultural purposes. A large drainage program along with an extensive clearing of forests gave rise to an extra 470 km^2 of arable land and increased the harvest of grain by a factor of about five during the nineteenth century. In 1880, the population had reached 55,000, but then several years of failing crops forced about 10 per cent of the population to emigrate from the island. Far from all drainage enterprises had been successful. The lowering of the groundwater level in the mires caused oxidation, rapid shrinking and decrease of the water-holding capacity of the organic peat soils, which made crops more sensitive to summer droughts (Folke, Chapter 8 of this book).

Also, manure was in short supply. It has been speculated that the bad crops in European agriculture in the late 19th century were partly due to insufficient recycling of soil nutrients to the fields (Pfister, 1990). Between 1850 and 1900, Gotland's agrarian system was directed towards producing grain for export while the number of cattle per unit area of arable land decreased by half (Table 2). In 1900, about 30 per cent of the

Table 2. Energy use, population growth, arable land and other ecological economic measures for the island of Gotland.

Measure			**Year**		
	1800	**1850**	**1900**	**1950**	**1980**
Population x 10^3	31	45	53	59	55
Rural/urban	7.5	9.7	5.3	1.6	0.9
Arable land, km2	150	244	650	830	830
ha arable land/capita	0.5	0.5	1.2	1.4	1.5
Cattle x 10^3	20	19	25	43	57
Cattle/km2 arable land	133	78	38	52	67
Grain yield, tonnes/ha	1	1	1.2	1.9	2.0
Fertilizers, kg/ha	-	-	46	370	600
Firewood, PJ	1.8	2.6	2.6	-	0.3
Coal and oil, PJ	-	-	0.7	3.3	11
Electricity, PJ	-	-	-	0.3	1.9

crop production was exported while the import of fertilizers to Gotland had barely began.

Production for export also influenced the productivity of forests on Gotland. Rising prises on the European market increased the demand for timber, firewood and other forest products which created an over-exploitation of the island's forest resources. By the end of the 19th century the yearly cutting of trees was almost twice as large as the regrowth, which gave rise to much local concern. Restrictions on cutting and prescription of tree planting were among the first environmental questions taken up by Swedish law. But, as the import of coal gained impetus during the first decades of the 20th century the demand for firewood dropped drastically. One may say that society's switch to fossil energies made possible a reestablishment of the forest productivity and a regrowth of the tree biomass.

5. THE TRANSITION TO FOSSIL ENERGIES

The import of coal made possible an expansion of the industrial and urban sectors, which in turn increased the demand for fuel and electricity. As new jobs became available in the island capital Visby, due to the increased trade and industrial activities, people began to move in from the countryside. But the industrialization process was rather slow in Gotland compared to the rest of Sweden. A major limiting factor was the restricted supply of electricity. A great deal of the imported coal was used for the production of electricity in local power-plants.

The increased use of fossil energies (see Figure 1) led to increased exploitation of the island's biophysical resources of land and water. The expansion of the agricultural sector continued during the first half of the 20th century, especially in the form of intensified cattle-breeding. The transformation of 200 km^2 of natural grazing-land and deciduous meadows to cultivated pastures implied a radical change in the rural landscape. But still the local production of animal feed did not meet the increasing demand. As the number of cattle doubled, the import of fodder increased by a factor of three. Consequently the self-sufficient type of agriculture which had prevailed for centuries in Gotland changed to an open type dependent on the import of fertilizers and fodder, influenced by the food prices on the Swedish and world markets.

The net income from the export of meat, dairy products and grain helped to fortify Gotland's financial position during the first decades of the 20th century. In 1939, the yearly import of fossil energy to Gotland had reached 4 PJ, but it then suddenly dropped by one half at the outbreak of World War II. Coal had to be strictly rationed and Gotland was forced to return to the locally available energy sources such as wood and peat. Most energy was used to ensure the supply of electricity and heating to households while industry and agriculture required only small amounts.

6. ECONOMIC GROWTH, EXTERNAL DEPENDENCY AND ENVIRONMENTAL CONSTRAINTS

During the first two decades after World War II, Gotland lost much of its former self-sufficiency. The fast transition from coal to oil and the new access to high-quality electricity by means of a cable from the Swedish mainland, made Gotland's economy almost totally dependent on imports. The imports of fertilizers quadrupled and made

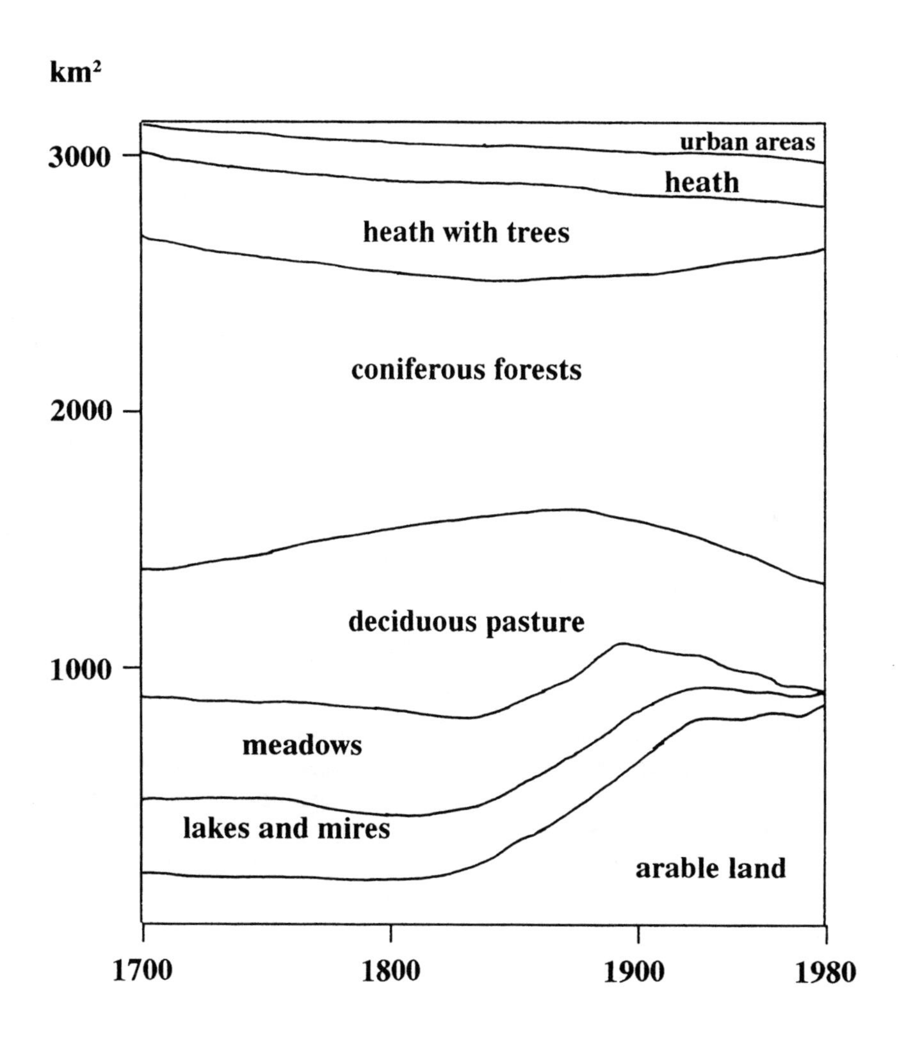

Figure 2. Landscape transformations in the island of Gotland 1700-1980.

it possible to double the total crop yield between 1945 and 1970 although the arable land area remained almost constant at about 810 km^2. At the end of this period, Gotland's agriculture was capable of producing protein to meet the needs of six times as many people as the resident population of about 55,000. But the intensification of agricultural production now seems to have reached its upper limit. The low soil moisture and the leaching of nitrates to the groundwater have become the main constraints on increasing agricultural production as well as other economic activities in Gotland. One sector which has grown significantly in the last few decades is the tourist industry which is now competing with agriculture for water. About 350,000 visitors to the island in the summer months make a large demand for water during a time of the year when there is a deficit of water in the hydrological regime.

The environmental history of Gotland shows the same general features as those described for other north-European areas (Brimblecombe & Pfister, 1990). The transformation of the regional landscape has been going on ever since Man came to Gotland some 7,000 years ago. The areal transformation during the three last centuries is shown in Figure 2.

This historical analysis confirms the notion that development occurs in pulses and with transitions to higher quality and more abundant energy sources as stated by Cottrell (1955), Odum (1973, 1983) and several others. The early development of agriculture and animal husbandry was linked to the temporal cycles of the terrestrial ecosystems, while in the last centuries the rapid growth of the human population and exploitation of natural resources was mainly associated with the use of imported energy resources. Cultural influences have also been of great importance.

The analysis reveals a pulsing pattern of prosperous growth and decay. During periods of prosperous growth there has been a tendency not to recognize the necessary support from Gotland's own ecosystems. Instead, imports were often used to increase the productivity and exploitation of the island's ecosystems, or to transform them into other uses dependent on foreign energy, resources and technologies.

The reduction in the biophysical support to Gotland's economic system during the last two hundred years is summarized in Table 3, which compares the photosynthetic energy-flows of the terrestrial ecosystems. The overexploitation of the forest biomass in the 18th and 19th centuries led to a 25 per cent decrease in the production of the coniferous forests. But as the imports of coal, and later of oil and electricity, replaced the use of wood fuel, the productivity of forests recovered and there is a current build up of the forest biomass of about 1.9 PJ per year.

The transformation of mires led to a reduction in the carrying capacity due to soil degradation and declines in water quality and quantity. It has been temporarily possible to enhance the carrying capacity of the region by increasing the agricultural productivity by the use of fertilizers and large inputs of other fossil energies (Table 3, Figure 2). In 1972, the subsidy of fossil energies to agriculture was 30 times larger than the solar energy contributions, expressed in units of similar energy quality. However, if these extra non-indigenous inputs were no longer available, the carrying capacity would decrease until the soils and water resources were restored which could take

Table 3. Gross primary production (solar fixation) in terrestrial ecosystems in the island of Gotland, and the use of industrial energy in agriculture (PJ solar equivalent per year).

Ecosystem	**Year**			
	1700	**1800**	**1900**	**1980**
Coniferous forests	43.8	35.1	33.3	39.3
Deciduous pasture	12.9	17.7	16.0	13.9
Meadows	7.7	9.4	5.0	0.7
Mires	16.3	16.3	11.6	2.1
Lakes	0.6	0.6	0.5	0.3
Heath with trees	3.2	2.6	2.6	1.4
Heath	0.2	0.3	0.5	1.1
Arable land	1.9	1.9	6.5	24.6
Total gross primary production	86.8	83.8	75.9	83.4
Industrial energy use in agriculture[a]	-	-	0.7	9.8

Note to Table 3.

a. One joule of industrial energy corresponds to about seven joules of gross primary production (Odum, 1983).

hundreds or even thousands of years. Furthermore, employing external energies in the transformation of forests, as happened in Gotland in the nineteenth century, could lead to a diminution in the entire carrying capacity of the island due to excessive soil erosion and declines in water quality and quantity.

Our investigations on the changing biophysical support to Gotland's economy seems to indicate that a better balance must be achieved between forests, wetlands, pastures and croplands in order to restore the island's capacity to supply society with appropriate resources of water, food, forestry products and fuels.

In addition, the potentials of the renewable energy of wind and waves should be more closely considered for generating electricity to prepare for the situation when outside support with power from the Swedish mainland comes to an end as the nuclear power is discontinued. Our studies indicate that with an increasing price of coal and oil these energy sources could supply all the electricity used on the island in 1975. When it comes to fuel energies, however, the restricted supply of freshwater becomes critical. To restore some of Gotland's wetlands for purposes of water conservation may be one of the most important ecological economic strategies to increase the necessary services available from the biophysical resource base to future generations of inhabitants on the island.

8. CONCLUSIONS

The regional study of landscape transformations, energy use and economic development in the island of Gotland, serves as a micro-scale example of global ecological economic interdependencies, and seems to illustrate a general pattern of prosperous growth and decay in Man-Nature relations.

Although it is not easy to establish a definite rule with regard to the issue of carrying capacity, because of the mix of resource, economic, environmental, social, and cultural factors that interact, it is still obvious that there is an ultimate limit on economic growth set by the biophysical resource base - the life-supporting environment. Different areas have different environmental limits to confront. For many countries and regions with rapid population growth or high rates of industrial development the potential risk of exceeding the carrying capacity is much more serious than in Gotland. This makes a systems view of economies and the environment even more relevant. In trying to achieve a viable balance between development and environmental protection, the systems ecology approach can serve as an important template for the incorporation of the ecological perspective into the traditional view of economic development. It is necessary that the perspective on economic development is broadening to include considerations of natural resource conservation and management of the life-supporting ecosystems to provide for the long-term viability of economies and their environments.

REFERENCES

Arrhenius, O. 1955. The Iron Age Settlements on Gotland and the Nature of the Soil: Vallhagar, A Migration Period Settlement on Gotland/Sweden. Köpenhamn, Stockholm.

Ayres, R. 1978. Resources, Environment, and Economics: Applications of the Materials/Energy Balance Principle. Wiley-Interscience, New York.

Bolin, B. 1986. How much CO_2 will remain in the Atmosphere? In: Bolin, B., Döös, B., Warrick, R. and Jäger, J. (eds.). The Greenhouse Effect: Climatic Change and Ecosystems. John Wiley and Sons, Chichester. pp. 93-155.

Bolin, B. 1990. Changing Climates. In: Friday, L. and Laskey, R. (eds.). The Fragile Environment. Cambrifge University Press, Cambridge. pp. 127-147.

Brimblecombe, P. and Pfister, C. (eds.). 1990. The Silent Countdown: Essays in European Environmental History. Springer-Verlag, Heidelberg.

Carlsson, D. 1979. The Development of the Cultural Landscape on Gotland: A Study of Changes of Agriculture and Settlements During the Iron Age. Liber Förlag, Visby, Sweden. (in Swedish).

Cleveland, C.J. 1987. Biophysical Economics: Historical Perspective and Current Research Trends. *Ecological Modelling* 38:47-73.

Cottrell, W.F. 1955. Energy and Society. McGraw-Hill, New York.

Ersson, P.G. 1974. Colonization and Devastation on Gotland. Meddelande från Kulturgeografiska Institutet, Stockholm University, Stockholm. (in Swedish).

Jansson, A.M. 1985. Natural Productivity and Regional Carrying Capacity for Human Activities on the Island of Gotland, Sweden. In: Hall, D.O., Myers, N. and Margaris, N.S. (eds.). Economics of Ecosystems Management. Dr. W. Junk Publishers, Dordrecht. pp. 85-91.

Jansson, A.M. and Zucchetto, J. 1978a. Energy, Economic and Ecological Relationships for Gotland, Sweden: A Regional Systems Study. *Ecological Bulletins* 28:154 pp.

Jansson, A.M. and Zucchetto, J. 1978b. Man, Nature and Energy Flow on the Island of Gotland, Sweden. *Ambio* 7:140-149.

Martinez-Alier, J. 1987. Ecological Economics. Basil Blackwell, Oxford.

Moberg, I. 1938. Gotland A.D. 1700: A Cultural Geographical Map Analysis. *Geogr. Ann.* 1-2:1-112.

Odum, H.T. 1971. Environment, Power, and Society. Wiley-Interscience, New York.

Odum, H.T. 1973. Energy, Ecology, and Economics. *Ambio* 2:220-227.

Odum, H.T. 1982. Pulsing, Power and Hierarchy. In: Mitsch, W.J., Ragade, R.K., Bosserman, R.W. and Dillon, J.A. (eds.). Energetics and Systems. Ann Arbor Science, Ann Arbor. pp. 33-59.

Odum, H.T. 1983. Systems Ecology: An Introduction. John Wiley & Sons, New York.

Odum, H.T. 1988. Living with Complexity: In: The Crafoord Prize in Biosciences 1987. Crafoord Lectures, The Royal Swedsih Academy of Sciences, Stockholm, Sweden. pp. 19-87.

Odum, H.T. and Odum, E.C. 1976. Energy Basis for Man and Nature. McGraw-Hill, New York.

Pearson, P. (ed.). 1989. Energy Policies in an Uncertain World. Macmillan Press, Hampshire, England.

Pfister, C. 1990. The Early Loss of Ecological Stability in an Agrarian Region. In: Brimblecombe, P. and Pfister, C. (eds.). 1990. The Silent Countdown: Essays in European Environmental History. Springer-Verlag, Heidelberg. pp. 37-55.

Vitousek, P.M., Ehrlich, P.R., Ehrlich, A.H. and Matson, P.A. 1986. Human Appropriation of the Products of Photosynthesis. *BioScience* 36:368-373.

WCED, World Commission for Environment and Development. 1987. Our Common Future. Oxford University Press, Oxford.

Zucchetto, J. and Jansson, A.M. 1979. Total Energy Analysis of Gotland's Agriculture: A Northern Temperate Zone Case Study. *Agro-Ecosystems* 5:329-344.

Zucchetto, J. and Jansson, A.M. 1985. Resources and Society: A Systems Ecology Study of the Island of Gotland, Sweden. Springer Verlag, Heidelberg.

Linking the Natural Environment and the Economy: Essays from the Eco-Eco Group,
Carl Folke and Tomas Kåberger (editors)
Second Edition.
1992. Kluwer Academic Publishers

CHAPTER 7

Arable Land as a Resource

by

Knut Per Hasund

Department of Economics
Swedish University of Agricultural Sciences
Box 7013, S-750 07 Uppsala, Sweden

Arable land can be treated as a production resource, with a natural resource component and an anthropogenic capital component. It is a stock, giving a renewable flow. The arable land resource has a quantitative dimension (hectares) and a qualitative one, which may be described by innumerable edaphic, topographic and climatic site parameters. Instead of measuring the resource stock (hectares, site parameter data), the resource is measured by its capacity to generate a flow. Physical measurements are derived by combining existing statistics with production functions. They are expressed in kilo barley-equivalents per hectare, and calculated for each of the 420 "homogenous" agricultural districts in Sweden. A concept called standard-hectares is developed, making acreage comparisons possible amongst different grades of land. The economic measure of the resources is land rent. As the residual of revenues minus costs in crop production it should reflect the different use-capacities of various plots of land. Swedish arable resources measured by land rents are fairly heterogeneous, showing distinct regional patterns. In 1983 the rent on Swedish tilled land was nearly normally distributed around a mean of US$ 100 per hectare. From 1968 to 1983 land rents declined, especially in the far south and in the north. The arable land resource situation is further illustrated by a new diagram that plots land rent against cumulative acreage. The barley-equivalents and land rents methods each provide a single, cardinal measure appropriate for comparing land of different quality. Further, a computerized model of Swedish arable resources was used to study the impacts of resource influencing factors. Possible resource situations are simulated on the basis of assumptions about the effects of factors like subsoil compaction, erosion, photochemical oxidants, on yields or costs. Air pollutants like ozone, subsoil compaction and urban expansion appear to be the most menacing factors. The possible effects are of such magnitude as to constitute a risk to the security of our food supply. Many of these factors are likely to exert their greatest negative influence on the plains in south and central Sweden, thus causing changes in the regional distribution of arable land resources, and also tending to make them less heterogeneous in the future.

1. INTRODUCTION

1.1 A Supply and Natural Resource Perspective

This investigation examines the arable land resources from a national supply and natural resource perspective. The aim is to develop a form from which to be able to deal with questions of the following nature: Will the arable resources be adequate in a situation where imports of food or inputs are not possible?[1] In case environmental or health causes restrict the use of fertilizers and biocides, how much agricultural production can be extracted from the land? What is the potential for producing raw materials for the chemical/technical industry or for energy purposes? What future privations will result from urban expansion on high yielding land? What risks to the land resources are constituted by resource damaging cultivation techniques and exogenous pollution?

In this paper we study land only as a resource for producing agricultural products. The varying capacities for producing externalities, such as landscape amenities, habitats or pollution, are thus not included. These aspects are, however, of the utmost importance, and the subject of further research.

First there is an analysis of the resource features and how they determine the choice of resource measures. The following section describes the methods developed for measuring the resource situation. It is presented together with some findings on Swedish arable land. Finally, possible future impacts on the resource are analyzed. The text is a selected extract, translated from Hasund (1986), with some additional material.

2. RESOURCE FEATURES

2.1 A Production Resource with a Renewable Flow

What characterizes arable land as a resource? In this paper it is claimed that it is not a pure natural resource, but a production resource, with a natural resource component and a man-made capital component. The natural resource component consists of the innate conditions: soil material, climate, topography and acreage of land. The capital component results from investments in land reclamation, drainage, fencing, soil improvements etc., that increase the productive capacity of the land. In due course, the kind of cultivation system used will influence the resource base as well. The capital and natural resource components together determine land's productive capacity, and they cannot be separated.

Arable land can be classified as a fund-resource:[2] a stock giving renewable flows of products. The flows consist of agricultural products in the form of raw materials for food, textiles, energy, etc. In principle, utilizing this flow does not affect the stock. The stock may, nevertheless, be intentionally or unintentionally increased or decreased by the cropping system, land investments, or exogeneous factors such as acidification, heavy metal deposits etc. These processes are more or less reversible, but mostly at a cost.

2.2 Determinants of the Resource

Resources only exist in relation to a certain society. To exist, the resource has to be demanded by society, and there must be sufficient knowledge to exploit it. A pre-requisite is of course the material basis, but if these additional conditions are not met, substances in nature are not resources, but mere phenomena. The term resource is used in many ways, but here we will use it for land that gives a positive return to society when exploited. Other land may constitute a potential resource, the land use-capacity of which may increase in the future so that it can provide a surplus of returns exceeding the cost of utilization if, for example, technology or prices change.

According to neoclassical economic theory, the magnitude of the arable land resource stock is determined by supply and demand. Since land is a production input, demand for arable land is a derived demand. It originates from the demand for agri-cultural products, and from the substitution possibilities regarding inputs as well as consumption. Input substitution refers to the farmers' inclination to use land relative to other inputs in the cultivation. It is determined by price and the marginal productiv-ity of land in relation to other inputs, such as fertilizers. Consumption substitution depends on price and quality of agricultural products compared to substitutes like fish or artificial fibres.

Technology, size of population, consumer preferences and incomes are essential in determining how much land will be considered viable. New technology for drainage in the 19th century suddenly made it feasible to turn huge areas of low- or non-yielding wetlands into some of the richest fields currently farmed in Sweden. Barbed wire is another example, an invention that drastically reduced the costs of using land for grazing. Without it, most of Swedish pastures would probably not be used for agricul-ture today. Another example can be found in Indonesia where terassing has made it possible to convert mountain slopes into rice fields. Without terrassing technology and population pressure, these slopes would not be agricultural land resources.

The remainder of this paper will concentrate on various supply attributes, demand and the technological aspects will not be considered further.

2.3 The Physical Supply

Arable land resources have a quantitative dimension (hectares), as well as a qualitative dimension (fertility). Fertility depends on a combination of edaphic, climatic and topographic factors. If the resource concept is also to reflect the qualitative dimension, it has to include all the factors that influence the land's production capacity. Consequently, the size of the arable land resource stock may be described by a set of site parameters, such as clay fraction at various soil depths, humus content, soil biota, cation exchange capacity, efficient soil depth, seasonal distribution of precipitation, length of growing season, etc. The number of more or less important site parameters, and possible combinations of them, is, in this huge complex of fertility factors, almost limitless.

Yield is determined by the crop's growth conditions, the growth factors. They are determined by the fertility factors in conjunction with the cultivation measures (see Figure 1). The difference between the concepts "growth factor" and "fertility factor" can be illustrated by this example: Crop-available water is a growth factor, partly determined by cultivation measures like irrigation and soil preparation, partly by the fertility factors precipitation, evaporation, groundwater level, capillarity, etc.

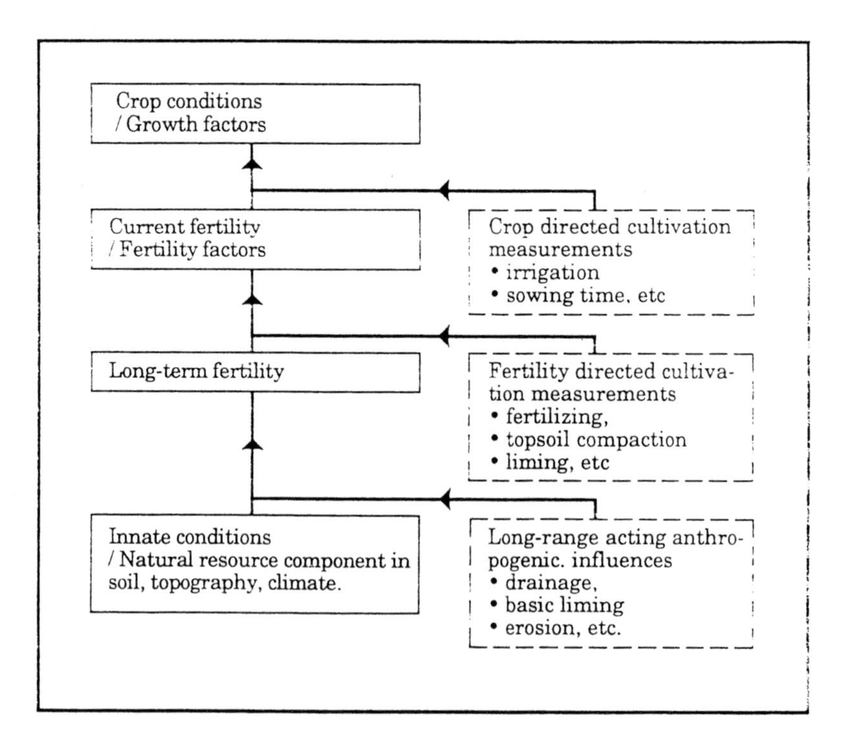

Figure 1. The temporal fertility levels of arable land.

The fertility directed cultivation measurements together with any unintentional influences comprise the anthropogenic component of the fertility factors. Their influence can be more or less enduring. For instance, fertilizing with potassium gives an effect that asymptotically decreases within a few years, while drainage generally improves the land's hydrological features for thirty years or more (Of course the gain of arable resources by draining have to be balanced against losses in the aquatic or other ecological systems).

Many edaphic fertility factors may be considered to consist of fractions along a time-scale. At one extremity on this scale is the fraction that influences short-term fertility, e.g. potassium in the soil solution. The other end also includes fractions that remain in a long-term view, e.g. potassium bound in mineral particles. The resource stock may thus be considered as fractions like russian dolls (a doll within a doll within a doll), where the longer term perspective includes less and less available or easily influenced fractions.

In this context it is important to make the distinction between the concepts "influenceability" and "permanence". Cultivation measurements and fertility factors may both be classified in these two dimensions (see Figure 2). The pH of the soil solution is an example of a fertility factor that it is easy or inexpensive to influence by liming, but the effect decreases rapidly. Whether fertility factors are easy or inexpensive to influence is not correlated to their permanence.

2.4 Economic Supply

The economic supply of arable land depends on the physical supply (the material base), institutional factors, the available technology etc. Unlike the land's physical use-capacity its economic use-capacity depends not only on fertility, but also on distance to markets, and on aspects of field layout such as size, shape and density.

Above, we noted that the physical supply consists of a natural resource component and a capital component. Draining, terracing, or building greenhouses would increase the supply when demand or new technique make these investments profitable. If land markets were perfect, land with the highest use-capacity would be cultivated first. Each increment of land supplied would incur higher costs of development, as well as higher cultivation costs per unit of production.

The possibilities to substitute other inputs for land are restricted by the law of diminishing marginal returns. We note however, that with a certain technology, agricultural production may be enhanced in two ways: by intensification (i.e. larger inputs of labour and other resources per unit of land), or by expanding the arable land stock. Increasing arable land resources can either be achieved by putting new land into production or by making investments that increase the productive capacity of existing fields.

When increasing agricultural production by intensification we are at the intensive margins of the arable land resources. This may be illustrated as a movement from the

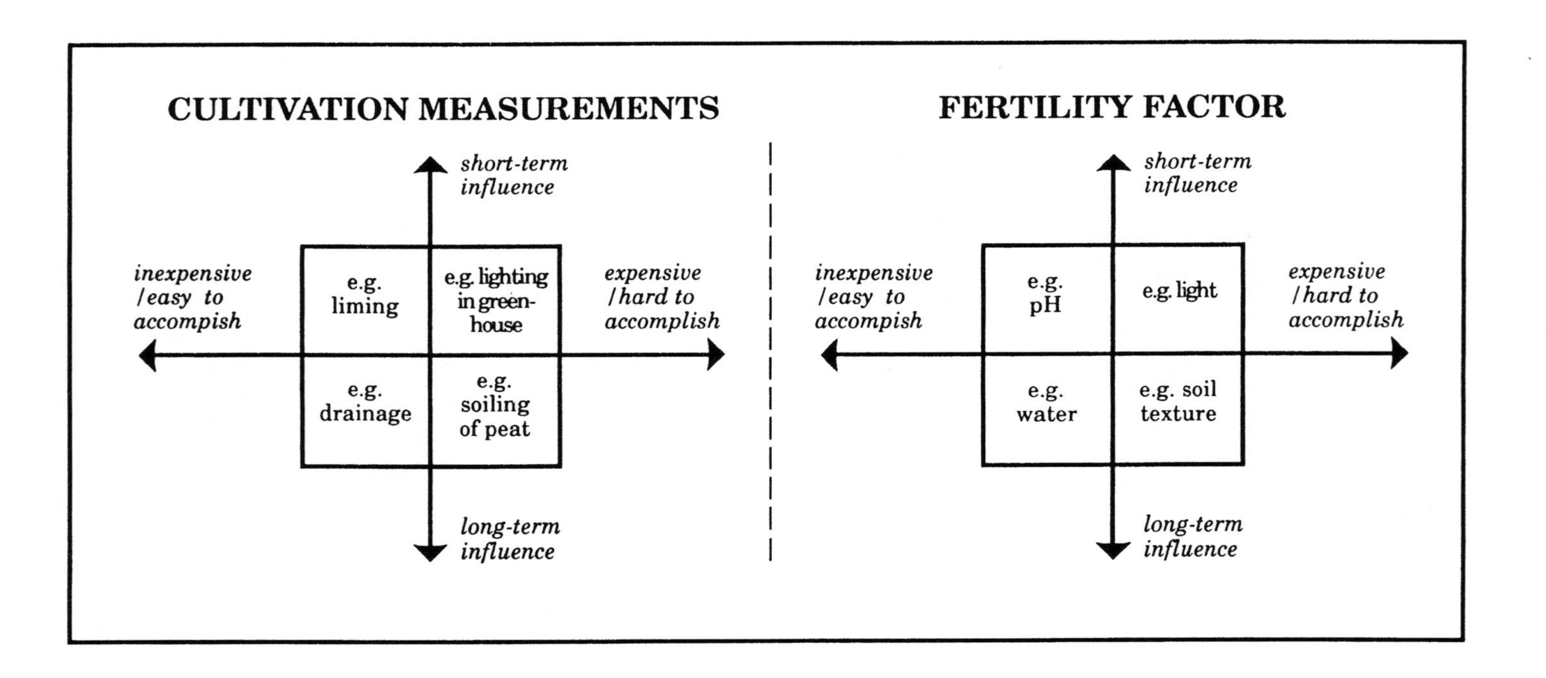

Figure 2. Illustration of how cultivation measurements and fertility factors can be classified according to influenceability and permanancy.

line MN to M"N" in Figure 3B. The intensive margin is defined as the point in the cultivation of a given piece of land at which the use of additional increments of labour and capital barely repays its cost (Barlowe, 1978).

The extensive margin corresponds to the lowest grade of land at which the cultivator may obtain sufficient production just to cover the non-land costs of production. In Figure 3, area C is the lowest grade of land that can be cultivated without losses (MVP=AVP=MC=AC), thus giving the extensive margin of land for a given output, price and technological level.

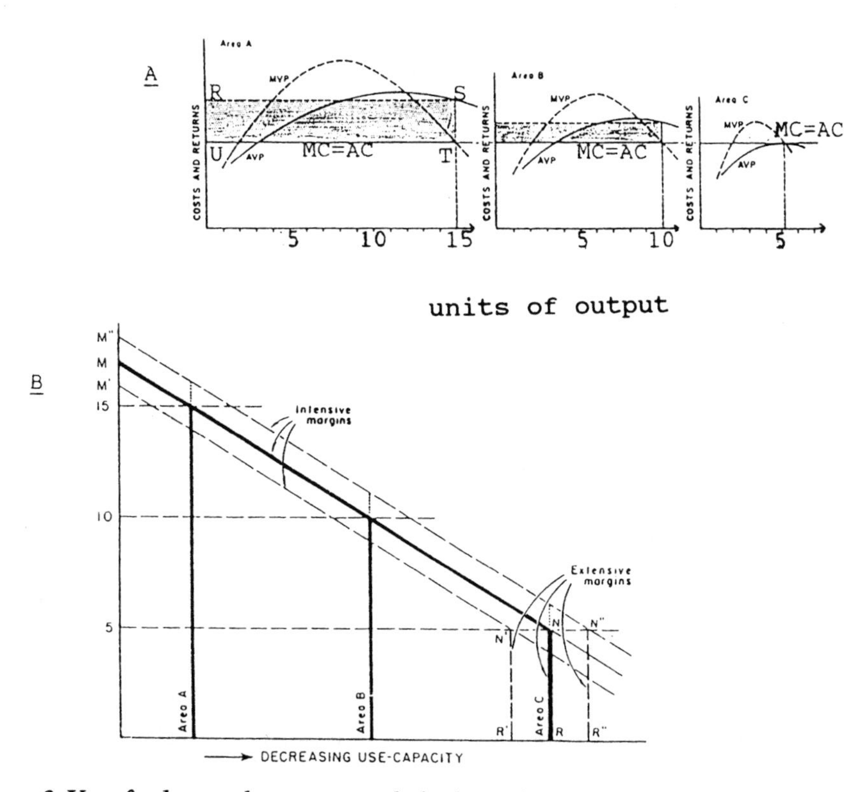

Figure 3. Use of value product curves to derive intensive and extensive margins of land use (Revised from Barlowe, 1978).

As can be seen from Figure 3, different grades of land have different MVP curves, and consequently, different optimal levels of intensity. In our example, MVP = MC at 15 units of input on land of grade A. At this intensity level, grade A land gives its maximum economic rent, corresponding to the shaded area RSTU in Figure 3A.

According to the theory of marginal productivity, the economic rent is the difference between revenues and costs of the production, i.e. q_i*(AVP-AC). The rent, in this case *land rent*, is treated as an economic surplus, a residual that is credited a factor of production. On land of grade B, the average cost (AC) is relatively higher when MVP = MC and the optimal level of production is lower, both circumstances implying a smaller land rent.

Investments that increase the productive capacity of existing fields, would in Figure 3 be described as raising the productive capacity of, for instance, some grade B land, into land of grade A, thus changing the shape of line MN.

The differences in fertility mean that arable land is not a homogeneous resource. This fact has an influence on how the resource should be treated in economic theory. One way is to standardize the fields and pastures, i.e. convert them into comparable units (cf. Table 1). Another way is to treat the resource as a number of distinct, interchangeable factor inputs, with different attributes, but among which substitution is possible.

3. MEASURING LAND CAPABILITY

3.1 Measuring the Flow

How can agricultural land resources be measured and described? Traditionally, they have been quantified by simple measures of the acreage of tilled land and pasture. To capture the qualitative dimension, one approach has been to classify land into a range of use-capacity categories. This method, however, does not allow for cardinal comparisons. It is not possible, for instance, to say how many hectares of class-4 land are equivalent to 10 hectares of class-1 land.

One improvement is to supplement the acreage measures with data on site parameters. However, this method is costly and also subject to several interpretational ambiguities (cf. Anon., 1976). For instance, soil texture which is optimal under one set of climatic and topographic conditions may not be so under another.

An alternative to stock measurements, is to measure resource assets by their flows: their capability to generate products in economic or physical terms. Such a method has been developed in several different forms, and applied to the agricultural land of Sweden.

3.2 Barley Equivalents as a Physical Measure

The Method

If the aim were merely to investigate the production capacity of fields, and to see how it is distributed throughout the country, physical measurements would be the most suitable. However, classifying land on the basis of its yield capacity in kilograms of protein or kilojoules of energy, has several shortcomings. For this reason, the land's capacity to yield barley was chosen as the variable on which to base the classification. Barley is a suitable reference crop, since it is *1)* cultivated all over the country, and *2)* representative (if barley yields are low, the land's use-capacity is generally low and vice versa).

The field classification is based on the barley *standard yields* for the 420 *yield survey districts* of Sweden (Anon., 1983a, b). These districts are demarcated to be as homogeneous as possible concerning the crop farming conditions: soils, topography, and climate. For crop insurance purposes, standard yields are calculated annually for each crop and district on the basis of several decades of empirical investigations. The standard yield is an estimate of the yield that can be expected if the weather and other conditions that influence the crops are quite normal (Anon., 1983a).

However, different cultivation techniques do influence the standard yields. To achieve a more pure measure of the land's use-capacity, the standard yields were calibrated to theoretical yields at 0 kg and 90 kg nitrogen fertilizer per hectare, using regional fertilizer statistics (Anon., 1979) and fertilizer-yield functions (Mattsson et al., 1981).

Results of the Physical Investigation

Swedish tilled land resources are approximately normally distributed around a median of 3,500 kilogram barley-equivalents per hectare (kg be/ha), at a fertilization level of 90 kg N/ha. The survey districts range from 2,100 to 5,100 kg be/ha. At 0 kg N/ha the theoretical median yield is 2,100 kg be/ha.

Comparing the survey districts, there is little difference in their ratings at 0 kg N as compared with 90 kg N. However, the relative difference between the highest and the lowest yielding survey districts is considerably greater at 0 kg N compared to 90 kg N per hectare. The standard deviation is 560 kg be/ha for both fertilizer levels, corresponding to 27 per cent and 16 per cent of the relevant median. Natural disparities in fertility may thus partly be levelled out by modern farming techniques.

Figure 4 gives a picture of the Swedish physical resource situation, when the 420 survey districts have been aggregated into eight production areas. These areas are still "homogeneous" with regard to crop farming conditions, but of course they are much less so than the yield survey districts.

To answer the two questions "How to compare arable land in different parts of the country?," and "How large are the arable land resources, and where are they located?", a new concept called a *field index* was developed. It is measured in *standard-hectares*. The standard-hectare for a survey district is calculated on the basis of its theoretical yield in comparison with acreage-weighted national average theoretical yields. From Table 1 we can see that, for example, that afforestation of 10,000 hectares in the County of Kristianstad, entails an agricultural loss of 11,300 standard-hectares, or average

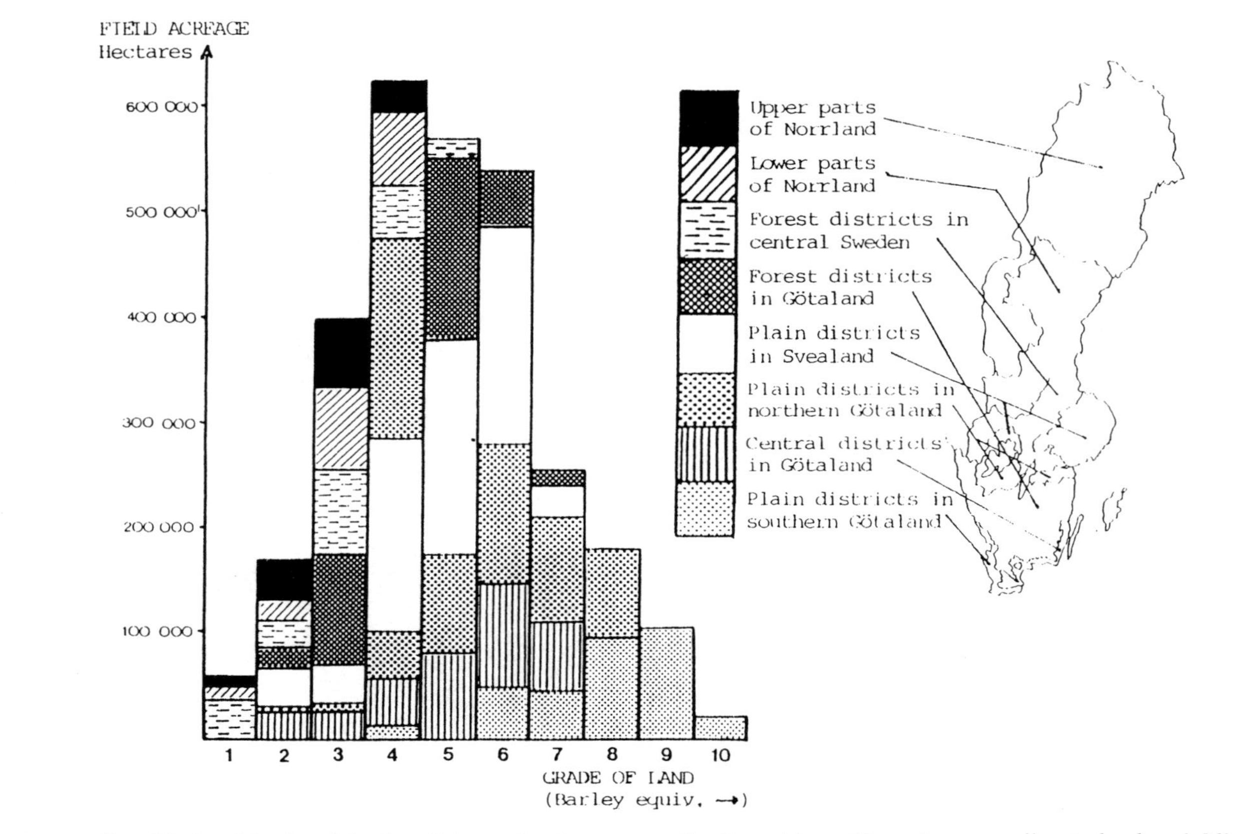

***Figure 4**. The acreage of arable land in the eight Swedish production areas, distributed into 10 grades according to barley yielding capacity at 90 kg N/ha 1983. Grade 1 has the lowest yield, less then 2,350 kg barley-equivalents per hectare. The grades are in steps of 300 kg be/ha.*

Swedish hectares. The differences in field indices will be greater when comparing the yield survey districts, as well as when comparing field indices based on theoretical yields at fertilization level of 0 kg N/ha, or based on economic yields.

The *arable resource values* can be expressed as the region's capacity to generate a flow of barley-equivalents at a fertilization level of 90 kg N/ha, i.e. theoretical yield multiplied by the field area. For example, Table 1 shows that the arable resource value of Malmöhus County is 1.10 million tonnes barley-equivalent. The county thus has approximately 13 per cent of the arable resources, but only 10 per cent of the Swedish field acreage.

Table 1. Arable land resources in some counties and yield survey districts in Sweden. Theoretical yields measured at fertilization level 90 kg N/ha. 1983. Field index for the whole country = 1.

	Field index	**Arable resource**	**Percentage of the total Swedish arable land resources**
	Standard-hectares	**Million tonnes barley-equival.**	**%**
<u>County</u>			
Uppsala	1.07	0.57	5.56
Malmöhus	1.31	1.10	13.08
Norrbotten	0.85	0.17	1.29
<u>Yield survey district</u>			
Highest field index	1.5		
Lowest field index	0.6		
The whole of Sweden	1.00	10.5	100.00

3.3 Land Rent as an Economic Measure

Arable land resources can also be measured by the capacity to generate a flow in economic terms. Land rent is such a measure.

Calculating Revenues minus Costs

In southern Sweden, revenues per hectare are substantially higher than in the north, but the cultivation is also encumbered with higher costs for pesticides, fertilizers, irrigation etc. To obtain a measure of the value of production for fields in different regions, land rent is chosen as the basis for the classification. As a residual of revenues minus costs, it reflects the various land use-capacities, differences that exist in soils, climate and topography.

Land rent has been calculated as the difference between revenues and incremental costs for the 16 main crops for each of the 420 "homogeneous" yield survey districts in Sweden. The calculations take into account the standard yields and acreage of each crop in the districts concerned. A standard yield value is then obtained by multiplying the yields by standard prices. The values have been taken from the Crop insurance statistics from the National Agricultural Marketing Board and from Statistics Sweden (Anon., 1974, 1983a, 1984a, b).

Data on costs are obtained from the regionalised gross margin data that are published annually by the Swedish University of Agricultural Sciences (SLU). These include all incremental costs that arise due to cultivation of a certain crop, such as seed, fertilizers, transport, drying, machines and labour (Anon., 1983c).

Based on land rents, all arable land was classified into 10 grades for illustrative purposes. The class width was 450 Swedish crowns[3] per hectare (SEK/ha). Arable land in survey districts with land rents higher than 2,500 SEK/ha is part of the highest class 10.

A Regional Pattern

Swedish arable resources measured by land rents are fairly heterogeneous (see Figure 5). In 1983 the rent on the total resources of tilled land was approximately normally distributed around a mean of 600 SEK/ha. The maximum was 2,700 SEK/ha. Distinct regional patterns exist. In general, land rents decrease from the south to the north, and from the plains to the woodland regions. This was expected, the investigation provided figures of the differences.

One reason land resources are more heterogeneous when measured in land rents than in physical terms (Figure 6) is that profitable crops have comparative advantages in some areas, but cannot be cultivated in other areas. Low grades of land may give reasonably good yields of energy or protein, but are hardly suitable for crops such as wheat or sugar-beet. Another reason is that costs do not fall as much as revenues do, when moving from a high to a lower yielding field. The difference in land rents is thus relatively larger than the difference in yields.

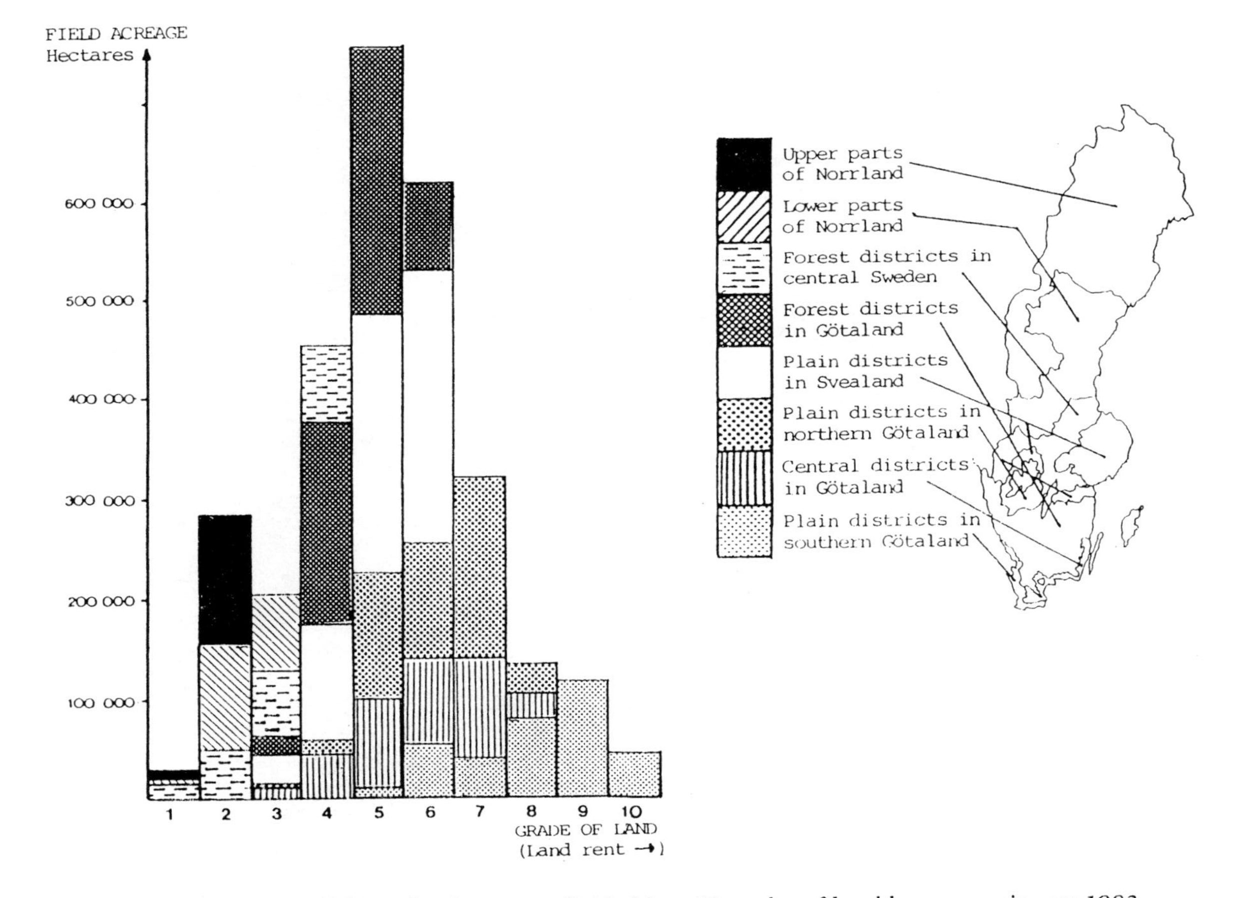

Figure 5. The arable land acreage of the Swedish production areas divided into 10 grades of land by economic rent 1983.

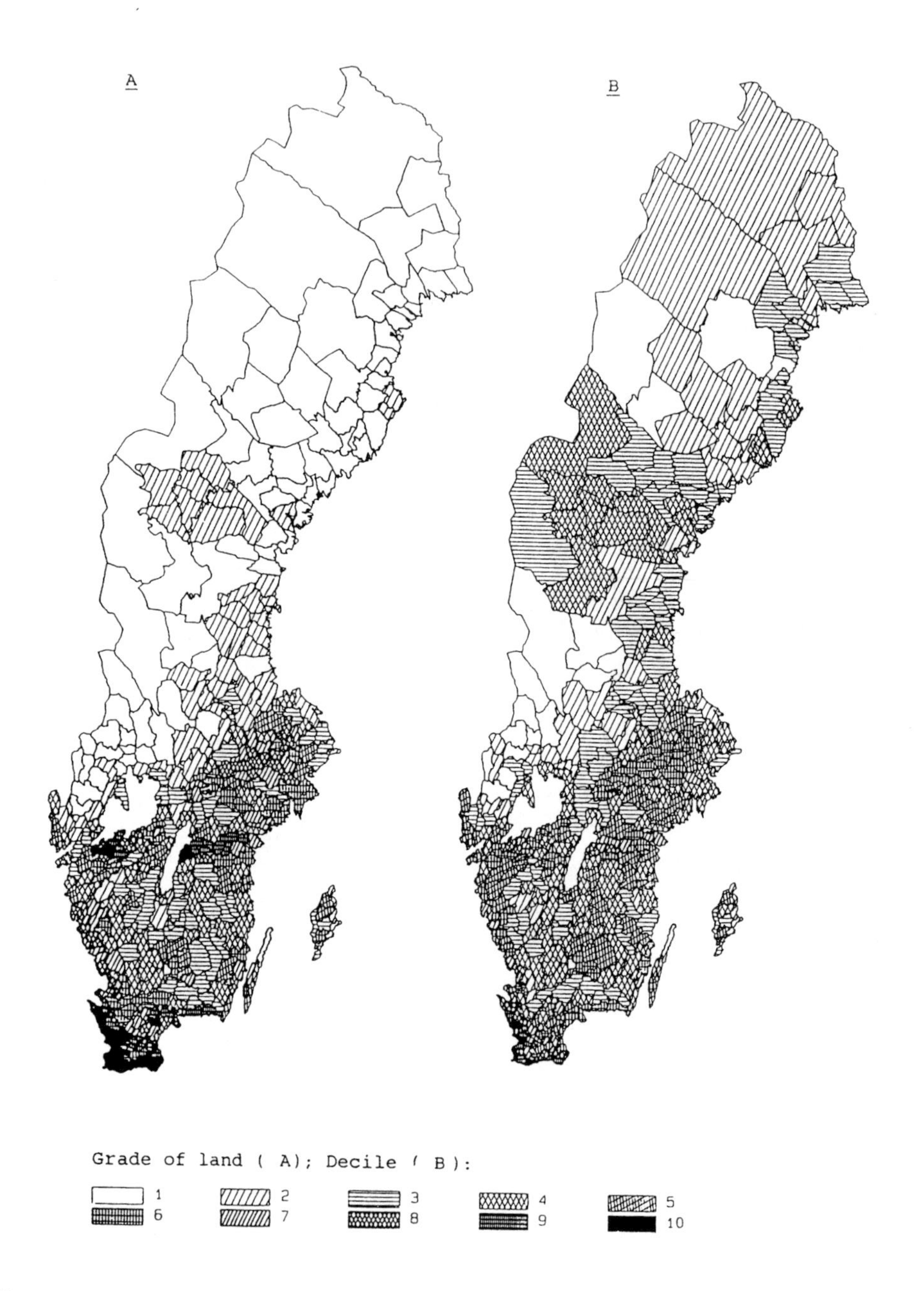

Figure 6. Grades of arable land in the 420 Swedish yield survey districts. 1983.

A: Classification by physical yielding capacity (barley-equivalents at 90 kg N/hectare, grades of land, grade 10 highest).

B: Classification by economic measures (land rent, deciles, decile 1 the lowest yielding 10 per cent of land).

The Land Rent Curve

A new diagramatic method has been developed to illustrate the arable land resource situation. It plots land rent measured in SEK/ha, against cumulative acreage (Figure 7). According to Ricardo and to later theories of marginal productivity, successively lower yielding grades of land are brought into cultivation when a larger production is required as demand rises. Inversely, the curve indicates how much land is threatened if the terms of trade for the agricultural sector, measured in land rents, decline.

In Figure 7 we can observe three segments. In the first, 0-0.5 Mha, land rents decline steeply from the highest yielding areas. In the middle, 0.5-2.3 Mha, the differences in rents are not particularly pronounced, and in the third segment 2.3-3.0 Mha there is again a sharp decline. The cause of the curve's steep slope at the lowest grades is most probably that the main part of the land with these properties has already been withdrawn from agriculture owing to low profitability.

Intertemporal Comparisons

The arable land resource stock is not immutable. Comparisons between the arable land situations of 1968/69 and 1983, indicate changes in total land use-capacity as well as in regional distribution. In general, land rents decreased. The plains in the south and middle of the country were *relatively* more valuable in 1983, having a smaller difference in land rents in relation to the superior land of the Plain districts of southern Götaland, and a superior position compared to the northern production areas.

The causes for the changes are *1)* some land is no longer used for agriculture because of low profitability, urban expansion etc., *2)* regionally heterogeneous investments in draining etc. that have increased the production capacity of the resource, *3)* technical innovations,[4] and *4)* relative price changes in agriculture.[5]

The Welfare Contribution of Arable Resources

Given the condition that the cost of all farm labour should be accounted at farm-worker salaries and that other inputs should be paid their market prices, it can be concluded that the area between the land rent curve and the field acreage axis corresponds to the concept "producer surplus". It is thus a minimum estimate of the land's contribution to national welfare. If desired, this surplus may be regarded as a gift from nature and the reclaimers to the land-owners or the national income. In 1983 the total net contribution was 1,700 MSEK (about US$ 300 Million). Table 2 shows the regional picture, when summing the land rents for every hectare in each production area.

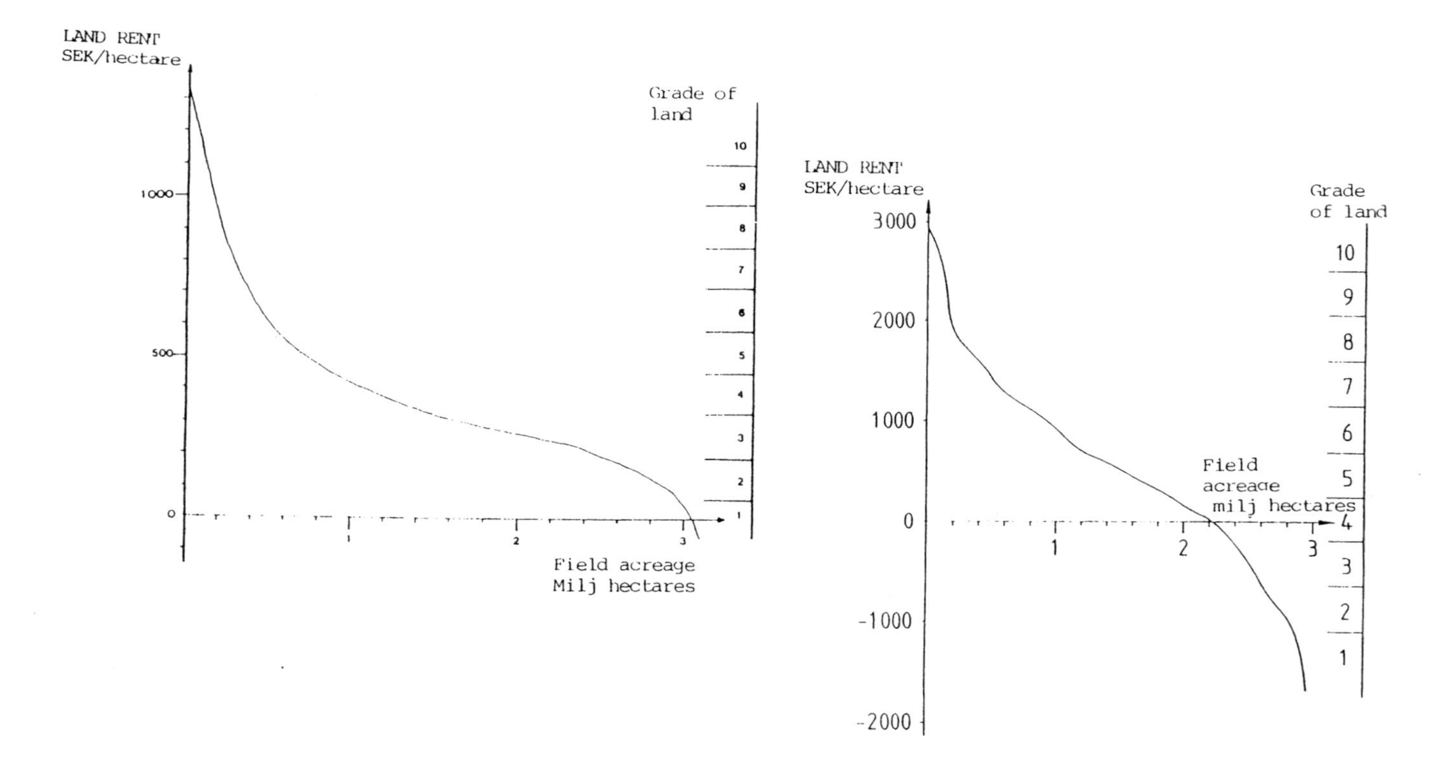

Figure 7. The Swedish arable land rent curves of the years 1968/69 and 1983. Value of money for respective year.

Table 2. Aggregated land rents of the Swedish production areas and the whole country, 1983.

Production area	**Positive land rents** **Million SEK**	**Negative land rents** **Million SEK**	**Total land rents** **Million SEK**
Plain districts in southern Götaland	655	0	655
Central districts in Götaland	311	-5	306
Plain districts in northern Götaland	480	-1	479
Plain districts in Svealand	402	-14	388
Forest districts in Götaland	207	-7	200
Forest districts in central Sweden	4	-96	-92
Lower parts of Norrland	0	-124	-124
Upper parts of Norrland	0	-135	-135
The whole of Sweden	2059	-383	1676

3.4 Pros and Cons of the Estimation Methods

Physical measures probably have greatest interest from a preparedness perspective, to calculate supply capability in scenarios of import restrictions, war, or increased demand and resource scarcity. This method also has the advantage of giving *a single* comparable measure, that is *more durable* than economic measures. In Sweden, it is also *cheap* and *fast* to perform since the required statistics are available.

The use of economic measures is based on the assumption that an optimal crop combination is used in each district, subject to the land's conditions for cultivation, prices, technology, crop sequence restrictions etc. The rent measure takes into consideration that some soils or climatic regions have a comparative advantage for cultivation of more valuable crops, although the differences in primary energy or protein yields are not particlularly large.

Land rent is a single, cardinal and monetary measure applicable everywhere. It is thus suitable for comparisons of land in different regions.

Land rent gives a fairly good picture of economic conditions for cultivating the land, but it is less appropriate for calculating the food producing capacity. The values obtained show the competitive powers of land in different soil or climatic conditions, and may identify in which areas there is a risk of farming being abandoned.

4. POSSIBILITIES AND THREATS TO ARABLE LAND RESOURCES

Which factors might influence Swedish arable land resources in the future? Historically, burn-beating, wind erosion, non-ecological cultivation systems, urban expansion etc., seriously damaged or destroyed substantial arable areas in the country. On the other hand, the resource has increased due to reclamation, soil formation, manuring, drainage etc. (water erosion, which internationally is a very large menace, is not considered a serious problem in Sweden).

If we use the model developed in section 2, arable land is both a naturally and an anthropogenically composed fund-resource. The stock gives a flow of products. It is possible to describe the stock by a number of fertility factors. Thus, if any of the edaphic, climatic or topographic factors are influenced, the resource stock as well as the flow will change.

The influencing factors may be extrinsic as well as intrinsic to agriculture. An example of an extrinsic influence is heavy-metal deposition due to industrial emissions in the atmosphere. In some cases, for instance sub-soil compaction, the influence has been due to ignorance, while in cases like urban expansion negative consequences for the resource are obvious (there has been a conscious encroachment).

The list of resource-influencing factors is long. Beside new and yet undiscovered factors, some factors are less important or empirically still little investigated. In this chapter, only supply influencing factors are considered. A more detailed statement is given for soil compaction and photochemical oxidants.

4.1 Soil Compaction

The mechanization of agriculture has brought about soil compaction by vehicles and machinery. Primarily, the volume of coarse soil pores decreases, which results in reduced permeability for air, water and roots. In the end, water saturated soils, anaerobic conditions, drainage difficulties, deteriorated timeliness etc., lead to more difficult or costly cultivation and reduced yields.

Causes

The damage increases with vehicle weight, the amount to which the vehicle is driven over each hectare, soil moisture and clay content, but is reduced by organic matter. It is thus possible to counteract compaction by the choice and design of machines and their wheels, planned (minimized) driving, time of driving, crop husbandry and drainage.

The pressure of a wheel on the soil surface is approximately equal to the tyre pressure. Pressure then decreases with soil depth, but much less gradually under a heavily loaded wheel than a lighter one with the same tyre pressure.

Consequently, for topsoil compaction, the tyre pressure is of major importance. In the subsoil, however, it is the total load, i.e. the axle weight, that determines the compaction. The critical limit for damage in the subsoil (>40 cm soil depth) is 6 tonnes axle weight. At 10 tonnes axle load, damage will occur below 50 cm (Håkansson, 1985b).

Axle loads greater than six tonnes were very unusual until ten or fifteen years ago, but now the number of heavy machines is increasing. Risks differ. Large tractors and machines are more common in the plains. The risks are greatest in sugar beet and potato cultivation, when raising power lines or using heavy grain carts, liquid manure spreaders, and lime trucks.

Influence on the Arable Resources

At present, top soil compaction reduces the yields by 3-4 per cent on average. Tillage, ground frost, drying and biological processes act to loosen the top soil layer, so, if compaction ceases, the damage disappears within a few years (Eriksson et al, 1975; Håkansson, 1985b).

To the extent that soil compaction entails a change in a fields' yield capacity, it will influence the size of arable resources. The topsoil compaction that has been taking place owing to agricultural mechanization involved a contraction of total resources by a few per cent. After transition to a cultivation system free from compaction, for instance using horse- or winch-pulled equipment, the yield capacity and resource would increase correspondingly within five or six years. Still more intensified topsoil compaction may be considered a temporary reduction of arable land resources.

Subsoil compaction, on the other hand, results in irreversible damage that accumulate over time. In the short run, the influence on yields is smaller than through top soil compaction. Ground frost, drying etc. have little or no rehabilitating effect, in this case. The possibility of loosening it by subsoilers is also in most cases discouraging: it is expensive, difficult and does not give lasting results. Thus, the deeper the compaction, the more permanent the damage (Håkansson, 1985b).

Calculating the Influence of Subsoil Compaction

Calculations have been made for two possible scenarios using the model of arable resources described in section 3. One scenario shows a probable resource situation if present trends are not broken. In the other scenario, the influence of subsoil compaction is calculated on the assumption that all arable land will be exposed sooner or later to loading by heavy vehicles.

The influence of subsoil compaction on arable land resources in physical terms is derived from the yield capacity in each of the 420 yield districts and multiplied by a reduction factor (<1). The reduction factor is specific to each district and determined by:

- The clay content of the dominant soil types in the district (Anon., 1953).
- The yield reducing effect (2 - 9 per cent) by subsoil compaction for different soil types (Håkansson, 1985a)
- The probability of subsoil compaction. In the first scenario, the risk depends on the crop pattern: the lowest risk, 0.3, for grains and oil-plants, the highest, 1.0, for sugar-beet (Håkansson, 1985a). Since the second scenario assumes that all land will be exposed, the probability of compaction is 1.0.

When calculating the resource measured in economic terms, i.e. land rent, it is assumed that the reduction of revenues due to subsoil compaction is proportional to the losses in yield, while costs are not influenced. Revenues are thus multiplied by the reduction factor of each district.

Subsoil Compaction Reduces Resources

Subsoil compaction gives rise to two kinds of natural resource impacts, each implying welfare losses. One impact is that the use of other resources, such as labour or fertilizers, has to be increased by input substitution to obtain a given output of products.

The other impact is that the yield capacity of the arable resources is reduced. The land rents of the Swedish arable resources will on an average decrease from 570 SEK/ha to 490 SEK/ha in the scenario of partial land compaction. Owing to the reduced yields, no land will be graded in the highest class 10. The largest losses are expected in the plains of the southern and central parts of the country.

The resource measured by its physical flow will not decrease as much as its economic flow. At the fertilization level 90 kg/ha, the average barley yielding capacity for the whole country will decrease from 3,560 kg to 3,490 kg barley equivalents per hectare. On the basis of existing experimental results and the adjusted assumptions of partial compaction, at the national level arable resources will thus decrease by about 2 pèr cent. Expressed in another way: subsoil compaction may bring about reductions of the yield capacity as large as if 56,000 standard hectares were irreversibly destroyed.

If, instead, it is assumed that all land will be exposed sooner or later to some vehicle with an axle load greater than 6 tonnes, the arable resources will probably decrease by 5 per cent, corresponding to 140,000 standard hectares.

Table 3. The impact of subsoil compaction on Swedish arable resources, measured as corresponding acreage losses.

Production area	Arable acreage	Qualitative acreage	Subsoil compaction impact Scenario 1	Subsoil compaction impact Scenario 2
	Hectares	Standard-hectares	Standard-hectares	Standard-hectares
Plain districts in southern Götaland	251,284	439,500	8,700	21,000
Central districts in Götaland	348,305	352,900	5,300	11,700
Plain districts in northern Götaland	470,481	514,100	11,100	31,400
Plain districts in Svealand	683,111	674,900	16,900	47,200
Forest districts in Götaland	558,931	525,400	7,600	17,100
Forest districts in central Sweden	209,733	171,100	2,900	6,600
Lower parts of Norrland	180,136	151,100	2,500	5,100
Upper parts of Norrland	142,635	115,600	1,400	2,700
The whole of Sweden	2,944,616	2,944,500	56,400	142,700

4.2 Photochemical Oxidants

In addition to acidification, which directly impairs land fertility, air pollution can damage crops so that yields are reduced. Air quality is thus a (climatic) fertility factor, and, in accordance with the approach to the arable resources developed in section 2, lower air quality implies a reduction of the arable resources.

Photochemical oxidants are in all probability the pollutants causing the most serious damage, among those which directly affect plants. The term refers to a large, heterogeneous group of oxidizing compounds that are formed in the air under the influence of solar radiation. In crop damage, ozone is by far the most important compound (Skärby, 1982, 1985a).

Cars and Other Sources

Ozone exists naturally in the stratosphere, where it is harmless and actually protects organisms from ultraviolet radiation. With lightning and strong atmospheric turbulence, increased concentrations may also occur in the troposphere where plants are growing. However, by far the greater part of tropospheric ozone originates from human activities.

Ozone is formed when nitrogen oxides react with hydrocarbons under the influence of solar radiation. The main source is auto emissions. More than 50 per cent of the nitrogen oxides, and a large part of the hydrocarbons come from traffic. Other combustion of fossil fuels and various industrial activities contribute most of the rest (Skärby, 1982).

Increasing Doses

The concentrations of ozone and other photochemical oxidants vary with time and between different sites. The temporal variations apply to:

1. Daily variations. Concentrations increase, especially on sunny days, throughout the morning, reach a peak in the afternoon, and then decrease in the evening. (Skärby, 1982)
2. Seasonal variations. Episodes of strongly increased concentrations over a limited time, as well as generally increased concentrations, occur in Scandinavia during the summer. Owing to climatic conditions, the number of episodes differs from one year to another.
3. Long-term trends. Ozone concentrations in the countryside in general (so-called "background levels"), have increased by 2-7 per cent annually since the 1950s. In Sweden, the background level normally varies between 30 and 80 $\mu g/m^3$. It is, however, not unusual for the concentration to rise to 200 - 300 $\mu g/m^3$, or more, during episodes of one hour to several days. For comparison, it might be mentioned that the recommended standard in Sweden for maximum concentra-

tion of ozone with regard to plant damage is 120 $\mu g/m^3$ during 1 hour per growing season, or 80 $\mu g/m^3$ on average for the season (Skärby, 1985a).

The highest concentrations occur in urban areas and 50-100 km downwind from them. In southern Sweden, the background levels are about the same as in western Europe, while they are lower in the north. Ozone can occur at a large distance from the source of pollution, and simultaneously over huge areas. It has been estimated that Swedish emissions of nitrogen oxides are about 300,000 tonnes a year, and that about the same amount is carried to the country by winds from Central Europe.

Impacts on Vegetation[6]

Plants absorb CO_2 and other gases, among them photochemical oxidants, through the stomata on their leaves. The first effects include increased permeability of cell membranes, and activation or inactivation of specific enzymes and hormones.

These molecular and cellular changes can impede important physiological processes, such as photosynthesis and the translocation of assimilation products within the plant. The result is, successively, reduced root growth, lowered reserves of hydrocarbons, visual leaf damage (cloroses or necroses), leaf shedding, reduced plant growth or even the death of the plant. From a cultivation point of view, there is an increased vulnerability to abiotic and biotic stress, reduced yields, and reduced quality of some products, especially leafy vegetables.

The magnitude of damage depends on species, variety, the plant's stage of development, environmental factors (e.g. SO_2, drought) and dose. Concerning the dose (concentration x time) the concentration carries more weight in the sense that a higher concentration during a short period results in more serious damage than the same dose distributed over a longer time of exposure.

Reduced Quality, Yields and Incomes

Crop quality may be impaired regarding the content of nutrients and other chemicals, taste, appearance, and hardiness in transport and storage. Danish experiments, for example, show that ozone decreases the nitrogen content of barley, and Swedish results reveal lower contents of protein and vitamin C in peas.

Negative yield responses have been verified in many reports, but there are still several unknowns in our knowledge of the quantitative impacts for several crops under Swedish conditions. A recent study (Hedvåg et al., 1990), using the best available knowledge, arrived at a value for the present losses to Swedish agriculture of an order of magnitude of 1,400 million SEK in a normal year.

In the USA it is estimated that the annual losses due to photochemical oxidants would decrease by US$ 1.9 billion (1982 dollars) if the ambient ozone were reduced by 25 per cent. An increase in the ozone concentration of the same magnitude would reduce yields by another US$ 2.1 billion (Adams et al., 1989). The situation is especially severe in California, where about 20 plant species cannot be cultivated at all due to the oxidants.

Impacts on Arable Resources

The assessment of present and future impacts on arable land resources is subject to major uncertainties, mainly due to the lack of knowledge of relevant dose-response functions, future pollution levels, and possible dynamic effects. The predominant prognoses predict increased concentrations in the next decades, in spite of all pollution control efforts. The main reasons are increased traffic and industrial activity in Sweden and Continental Europe, and the relative expense of reducing the emission of nitrogen oxides.

The difference between present background levels and highly phytotoxic concentrations are small. There is thus a high risk of seriously reduced yields and contraction of arable land resources. In recent years, plant breeding aimed especially towards ozone tolerant varieties has been accomplished, e.g. in Poland and the USA, but the outcome has not yet been demonstrated (Skärby, 1985a).

If, contrary to expectations, pollution were to decrease, it is most probable that the land's capacity to yield would increase several per cent in southern Sweden. If, on the other hand, prognoses and experiments are right, yield reductions of 20-30 per cent are not implausible in the foreseeable future (Skärby, 1982, 1985b).

To estimate the land resource situation in a plausible scenario of increased ozone content in the troposphere, a comparative static analysis has been carried out. Calculations are based on the models of the Swedish arable land presented in part 3, and the following assumptions on yield reductions are made: potatoes and sugar-beet -25 per cent, grain and oil plants -15 per cent, and ley -10 per cent (Skärby, 1985b). For northern Sweden it is assumed that yields will not be affected.

The simulation implies that land rents will decrease from roughly 600 to about zero SEK/ha (national average). Since it is southern Sweden that is most likely to be affected, the disparities in land rents between the regions will become smaller. Differences in crop patterns also tend towards levelling of land rents. Consequently, in this scenario, no land would yield enough to rank in any of the three highest economic classes. To summarize, arable land resources measured by land rents, will *a)* contract in total, *b)* shift regionally, and *c)* become less heterogeneous.

The arable land resources measured by barley equivalents will not contract as much as when measured in economic terms, since it is just the yields and not the costs that are influenced. Taking into account crop patterns, existing dose-response functions and pollution prognoses, the physical resource is calculated to diminish by approximately 10 per cent in total. That corresponds in yield capacity to the loss of 360,000 standard-hectares. However, substantially larger yield reductions may occur in the next decades, but it is also possible that the emissions will be reduced and damage counteracted.

4.3 A Comprehensive View of Resource Impacts

There are four main ways in which the arable land's value as a resource may be influenced, positively or negatively, as illustrated in Figure 8 below. First, the acreage of arable land may be reduced, e.g. by erosion or urban expansion. A second impact is that the resource measured by its flow or use-capacity may decrease due to lower hectare yields, caused by ozone, soil compaction, erosion, etc. Furthermore, several factors may increase the costs of cultivation, thus reducing land rents. Soil acidification, for instance, increases the cost of liming, and wind erosion may necessitate resowing (Nihlén, 1984). Finally, the quality of agricultural products often deteriorates owing to soil poisoning and ozone, this too reduces land rents or consumer welfare.

Factors induced by human activities completely dominate the picture of potential changes in arable land resources. Only a small fraction of the acidification or the erosion may be credited "natural" processes.

The potential to change the agricultural land resources differs between the factors. The probability of the various influences also differs. Figure 9 is an attempt to provide a composite illustration of how the Swedish arable land resources may be influenced within the next few decades.

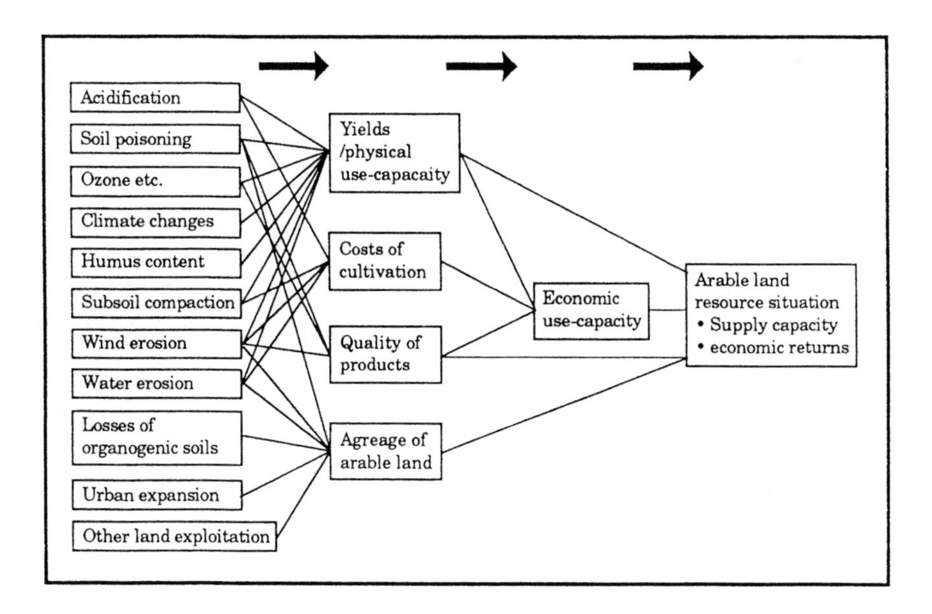

Figure 8. Relations between resource influencing factors and resource impacts.

From the figure it is evident that there is much uncertainty about future changes. The negative influences dominate the picture, thus the prospects do not make it reasonable to expect any *spontaneous* improvements. However, for all the factors discussed there are measures which could provide improvements. There are also possibilities to counteract the threats and compensate for the resource degradation or increased demands by means of land reclamation and land improvements.

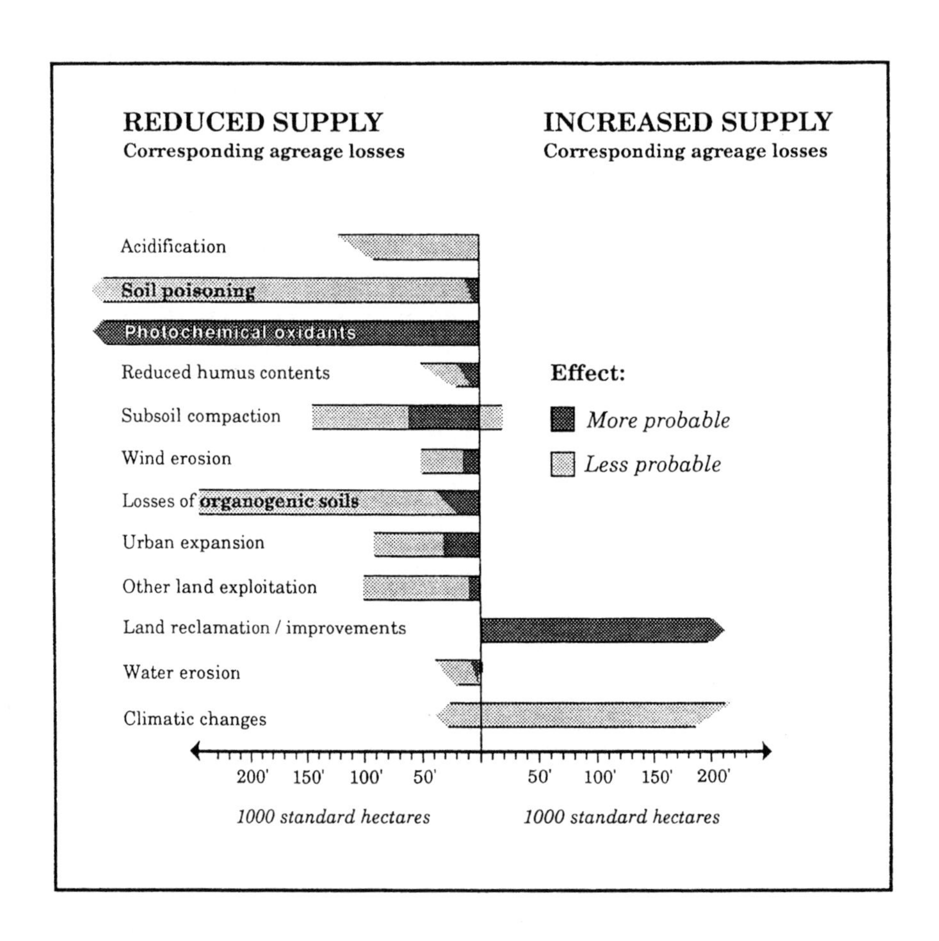

Figure 9. The potential and probability of some factors influencing Swedish arable land resources within the next 25 years, expressed in corresponding changes in standard-hectares. Prognosis 1985.

Air pollutants such as ozone seem to be the biggest threat to future arable resources. Soil compaction and losses of organogenic soils through cultivation may also seriously impair the yield capability, but such damage is possibly easier to avoid. It is highly possible that urban expansion and similar land exploitation will entail substantial irreversible losses of cultivated land. The impacts are to some extent additive, and in some cases synergistic. Taken together, the tendencies involve risks of rather large encroachments, damage that might correspond to more than half a million hectares. If the negative impacts become too large, the national capacity for food self-sufficiency would be in danger, especially so in case of other stresses. For example, in a situation of reduced land margins, limited use of fertilizers due to trade blockades or environmental reasons, might become a very painful strain.

When compiled, many factors tend to act in the direction of reducing the arable land resources of the plains in southern and central Sweden. Consequently, the land in the north and in the forest regions will become of greater relative importance. The causes are:

1. The location and the topography of the plains soils make them more exposed to air pollution, wind erosion, exploitation for urban uses, shale strip mining etc.
2. The crop farming systems in the plains, dominated by large scale production of nothing but grain and oil seed, involve a larger risk of soil poisoning by pesticides and sewage sludge, wind erosion, soil compaction and depressed humus content.
3. The soil types in the plains are often more sensitive to wind erosion or soil compaction.

The investigation indicates that it is in general the highest yielding land that runs the risk of being most negatively effected. This has an equalizing impact that probably makes the nation's arable land resources less heterogeneous, at least in their physical capacity.

5. CONCLUSION

Considering the very complex set of causes, it should be emphasized that economizing arable land resources can and should be carried out on several levels. The problems can and should be tackled on the farm level, the political level, the national level and internationally, as well as in research, technological development, education, and the formation of attitudes and opinions. On the political level, much can be gained if taxes, grants and restrictions are designed taking into consideration their effects on arable land resources. This does not only concern agricultural policy; macro policy and other sector policies such as energy, housing and environmental policies must also be adapted to the necessity of resource conservation.

Hopefully, this investigation of arable resources will be useful when shaping agricultural policies to promote a production system that uses techniques involving less pollution that involves a smaller consumption of finite resources, and that is based to a higher extent on sustainable use of the renewable resource land.

ACKNOWLEDGEMENTS

The author gratefully acknowledges the support and interest of David Vail and his help with improving the article, and constructive comments from the two referees Pierre Crosson and David Pimentel.

NOTES

1. Sweden is currently a net exporter of food, but to nearly 100 per cent dependent on imports of fertilizers and pesticides. Since 1948, one of three main agricultural policy objectives has been national food security.
2. Terminology refers to "funds" giving a perpetual return to be used, versus "inventory-resources" whose stock decreases when exploited.
3. 1 SEK is about US$ 0.16
4. An example is the new type of combine drill that came into use during the 1970s, increasing the resistance of the crops to drought, thus enhancing the value of the land resources especially in eastern Sweden. New varieties of spring oil-plants and wheat also substantially favour the plains in the middle of the country.
5. For instance, prices of oil-plants and potatoes for industrial use have increased more than the average product index, and of course this favours the regions where these crops can be cultivated.
6. Literature cited in the subsection: Hedvåg et al., 1990; Skärby, 1982, 1985a.

REFERENCES

Adams, R.M., Glyer, J.D., Johnson, S.L. and McCarl, B.A. (1989) . A reassessment of the economic effects of ozone on U.S. agriculture. *JAPCA* 39:960-968.

Anonymous, 1953. Atlas över Sverige 63-64. Åkermarkens matjordstyper. Svenska sällskapet för antropologi och geografi. Stockholm.

Anonymous. 1974. Statistiska meddelanden J 1974:20. Statistics Sweden, Stockholm, Sweden.

Anonymous. 1976. Den Danske Jordklassifisering. Teknisk redegörelse. Landbrugsministeriet, Copenhagen.

Anonymous. 1979. Statistiska meddelanden J 1979:8.6. Statistics Sweden, Stockholm, Sweden.

Anonymous. 1983a. Statistiska meddelanden J 1983:7.2. Statistics Sweden, Stockholm, Sweden.

Anonymous. 1983b. Statistiska meddelanden J 1983:8.1. Statistics Sweden, Stockholm, Sweden.

Anonymous. 1983c. Områdeskalkyler - Jordbruk 1983. Research Information Centre. The Swedish University of Agricultural Sciences, Uppsala.

Anonymous. 1984a. Statistiska meddelanden J 1984:8.1. Statistics Sweden, Stockholm, Sweden.

Anonymous. 1984b. Statistiska meddelanden J 1984:9.1. Statistics Sweden, Stockholm, Sweden.

Barlowe, R. 1978. Land Resource Economics. Prentice-Hall, Englewood Cliffs, New Jersey.

Dent, D. and Young, A. 1981. Soil Survey and Land Evaluation. George Allen & Unwin, London.

Eriksson, J., Håkansson, I.and Danfors, B. 1975. Jordpackning - Problem inom det Högmekaniserade Jordbruket. Konsulentavdelningens stencilserie. Allmänt 1. The Swedish University of Agricultural Sciences, Uppsala.

Food and Agriculture Organization of the United Nations. 1976. A Framework for Land Evaluation. *Soils Bulletin* 32.

Håkansson, I. 1985a. Personal communication. Department of Soil Sciences, The Swedish University of Agricultural Sciences, Uppsala.

Håkansson, I. 1985b. Swedish Experiments on Subsoil Compaction by Vehicles with High Axle Load. *Soil Use and Managent* 1985(4).

Hasund, K.P. 1986. Jordbruksmarken i Naturresursekonomiskt Perspektiv. Department of Economics, Report 269. Swedish University of Agricultural Sciences, Uppsala.

Hedvåg, L., Hasund, K.P. and Pleijel, H. 1990. Ekonomiska Konsekvenser av Ozonpåverkan på Jordbruksgrödor. Swedish Environmental Protection Agency, Stockholm. In press.

Mattsson, L. and Bjärsjö, J. 1981. Kvävegödsling till Korn. Department of Soil Sciences, Report 135. The Swedish University of Agricultural Sciences, Uppsala.

Nihlén, T. 1984. Utredning av den för Vinderosion Utsatta Marken i Skåne. Department of Physical Geography, Report 58. University of Lund, Lund.

Ricardo, D. 1817. On the Principles of Political Economy and Taxation. Murray, London.

Salter, L.A. 1967. A Critical Review of Research in Land Economics. The University of Wisconsin Press, Madison, Wisconsin.

Skärby, L. 1982. Effekter av Luftföroreningar på Vegetation. Fotokemiska Oxidanter. Swedish Environmental Protection Agency, PM 1562, Stockholm.

Skärby, L. 1985a. Effekter av Gasformiga Luftföroreningar på Vegetetion med särskild hänsyn till Fotokemiska Oxidanter. Report from the Institute for Water and Air Pollution Research (IVL), Gothenburg.

Skärby, L. 1985b. Personal communication. IVL, Gothenburg.

Linking the Natural Environment and the Economy; Essays from the Eco-Eco Group,
Carl Folke and Tomas Kåberger (editors)
Second Edition.
1992. Kluwer Academic Publishers

CHAPTER 8

The Societal Value of Wetland Life-Support

by

Carl Folke

Beijer International Institute of Ecological Economics
The Royal Swedish Academy of Sciences
Box 10005, S-104 05, Stockholm, Sweden
and Department of Systems Ecology, Stockholm University

In this chapter I analyze the societal value of a Swedish wetland system with respect to the various economic functions such as cleansing nutrients and pollutants, maintaining the level and quality of the drinking water, processing sewage, serving as a filter to coastal waters, sustaining genetic diversity and preserving endangered species. The major part of such life-support functions have been lost, due to extensive exploitation. I evaluate the loss in terms of the reduced solar energy fixing ability (Gross Primary Production, GPP) and the deterioration of the stored peat, and compare it to the cost of replacing these environmental functions with technical processes, estimated in monetary terms and industrial energy terms. Such substitutes include irrigation dams, water transportation, well-drilling, water purification, sewage treatment plants, fertilizers, fish farming, and efforts to save endangered species. I find that the undiscounted annual monetary replacement cost is of the order 2.5 to 7 million Swedish crowns (US$0.4-1.1 million), and that the annual industrial energy cost between 15 and 50 TJ approaches the annual loss of GPP of 55 to 75 TJ, when both are expressed in units of the same energy quality (fossil fuel equivalents). The major part of the technical replacements concerns biogeochemical processes and the hydrological cycle. Not more than about 10 per cent are related directly to the biological part of the wetland system. The present biophysical analysis serves as an indicator of the true life-support value of a wetland system, and is thus a useful complement to economic analysis. It is concluded that ecosystems perform a lot of valuable and necessary work at no cost, and that industrial technologies should be developed to supplement and enhance this support instead of having to replace it when it has already been destroyed.

1. INTRODUCTION

In Sweden, as in other parts of the world there has been a huge conversion and degradation of wetlands, mainly for agriculture and forestry, but also for peat-cutting and urban uses. Often wetlands have been regarded as nonproductive areas which through drainage, diking and filling could be transformed into more useful land. However, wetlands are very important ecosystems, providing human societies with essential and highly valuable life-supporting functions (Odum, 1989). Monetary valuations of wetlands have usually concentrated on specific environmental services such as water supply, flood prevention, pollution reduction, fish and wildlife production, the provision of recreation and aesthetics (e.g. Hammack & Brown, 1974; Lynne et al., 1981; Thibodeau & Ostro, 1981; Batie & Shabman, 1982; Faber, 1987). Less frequently, attempts have been made to estimate the life-support value of entire wetland ecosystems, which are generally required to sustain each environmental service (e.g. Gosselink et al., 1974; Odum, 1984; Costanza et al., 1989).

This chapter is an attempt to quantify the life-support value of a Swedish wetland system, the Martebomire, on the island of Gotland in the Baltic Sea. This mire has been subject to extensive draining, starting around 1850, and most of the wetland-derived environmental goods and services have been lost. The purpose of the analysis is to compare the loss of the wetland's support functions with the costs for replacing them with technical solutions, to the extent that this is possible. I do not estimate the benefits of the wetland exploitation, since the objective is to illustrate the value of the wetland's life-support, not of its alternative uses, such as intensive agriculture.

It is difficult to analyze the significance of ecosystem life-support solely in monetary terms, because environmental functions seldom have a market and the general public seldom has perfect information about the role of ecosystems in sustaining human activities. The term "value" in this chapter does not refer to human preferences based monetary valuation, but to the wetland's functional role for society - the life-support. The life-supporting ecosystems provide the physiological necessities of life by a complex network of living and non-living parts of the environment dynamically operating on different time-scales (Odum, 1989).

Another approach to the evaluation of the life-support provided by ecosystems is to trace the flows of energy and materials[1] required to produce and maintain goods and services, whether environmental or man-made. An energy analysis is one approach for comparing the work of "nature" with the work of "the economy", and which makes it possible to consider environmental functions which seldom have a market and on which the general public seldom have perfect information. This is not to argue for an energy theory of value. The use of energy is a necessary but not sufficient condition for things to have a value. The approach is useful as a complement to economic analysis, which is discussed at the end of this chapter.

I evaluate the loss of the wetland's support functions in terms of the reduced embodied solar energy[2] and compare it with the embodied industrial energy[2] required to produce and maintain the technical substitutes. I also estimate the monetary replacement costs for these substitutes. The paper illustrates that economic activities not only effect the environment, but also require often unacknowledged life-support functions of ecosystems.

1.1 Wetlands Support Society

Wetlands are ecological interfaces or transition zones between terrestrial uplands and deep water aquatic systems. There is a large diversity of wetlands (Mitsch & Gosselink, 1986; Day et al., 1989), and some of them are among the most productive ecosystems on earth (Whittaker & Likens, 1975). Hydrological pathways, which transport energy and nutrients to and from wetlands, are fundamental for the establishment, and maintenance of these ecosystems and their processes. The hydrological regime influences pH, oxygen availability, and nutrient flux, and these parameters to a large extent determine development of biota. The biotic part of the ecosystem in turn modifies the hydrological conditions by trapping sediments, interrupting water flows, and building peat deposits (Gosselink & Turner, 1978) (Figure 1). Wetlands in general have low oxygen levels in the water column, reducing conditions in sediments, and a complex chemical cycling. When streams enter wetlands the water velocity is reduced causing sedimentation and the chemicals sorbed to sediments to be deposited in the wetland. There is significant sediment-water exchange due to the shallow water in wetlands, and a diversity of decomposing organisms and decomposition processes in wetland sediments lead to processes such as denitrification and chemical precipitation which remove chemicals from the water. The high productivity rate of many wetlands often implies high rates of mineral uptake by vegetation and subsequent burial in sediments when the plants die. The accumulation of organic materials causes the burial of chemicals including toxic substances (Mitsch & Gosselink, 1986).

Wetlands are valuable for many different reasons. They change sharp run-off peaks, from heavy rains and storms, to slower discharges over longer periods of time (Odum, 1981), and thus prevent floods. They recharge groundwater aquifers, and provide drinking water directly in dry seasons. Many wetlands act as sinks for inorganic nutrients and many are sources of organic materials to downstream or adjacent ecosystems. They improve water quality and often serve as a filter for wastes, reducing erosion and the transport of nutrients and organic materials, sediments and toxic substances into coastal areas (Stearns, 1978; Mitsch & Gosselink, 1986).

Wetlands are important habitats and serve as nursery and feeding areas for both aquatic and terrestrial migratory species. In fact, the abundance of wetlands in estuarine regions is often correlated with regional fish catch (Day et al., 1989). Wetlands generally have high recreational and aesthetic values. Such values would, however, not be sustained without the life-support functions. Furthermore, wetlands are involved in global biogeochemical cycles and contribute to the global stability of available nitrogen, atmospheric sulfur, carbon dioxide, and methane. In other words, wetlands provide an abundance of environmental services of high value to society

(Turner, 1988). It is, however, important to remember that these services cannot be managed as independent commodities, as they require the support of the wetland's processes and functions to be produced and maintained. Each is the result of the work of the whole wetland system[3] (Figure 1).

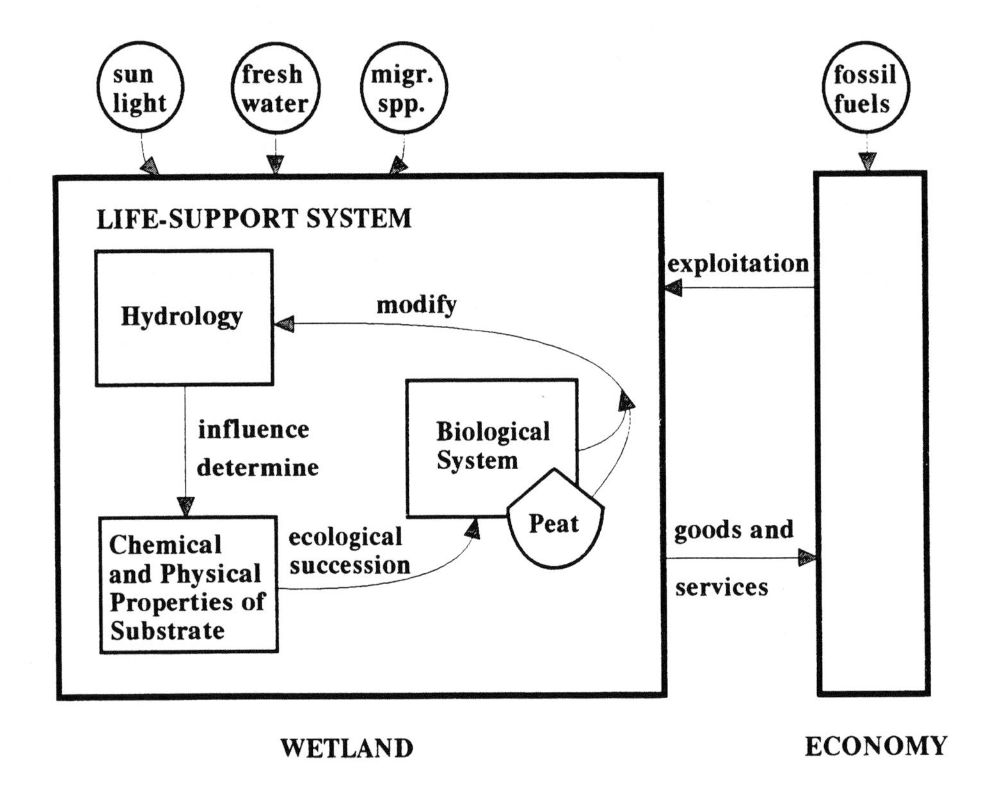

Figure 1. Conceptual model of the symbiosis between hydrology, chemical parameters, physical properties of substrate, and biota of a wetland ecosystem forming the life-support which sustains environmental goods and services used by humans (modified from Gosselink & Turner, 1978).

2. MATERIAL AND METHODS

2.1 The Wetland Martebomire

In the middle of the nineteenth century about 10 per cent (300 km^2) of the Swedish island of Gotland was covered by mires. Due to extensive drainage activities, mainly for agricultural purposes, about 90 per cent of these freshwater wetlands have disappeared (Figure 2). The Gotland mires are high-productive fens[4] developed in shallow basins of the limestone bedrock (Figure 3). They are dominated by *Cladium mariscus* stands, *Carex* species, *Salix* shrubs (Zucchetto & Jansson, 1985) and *Phragmites communis* stands (Petterson, 1946) with only minor elements of *Sphagnum* species (von Post, 1929; Lundqvist et al., 1940).

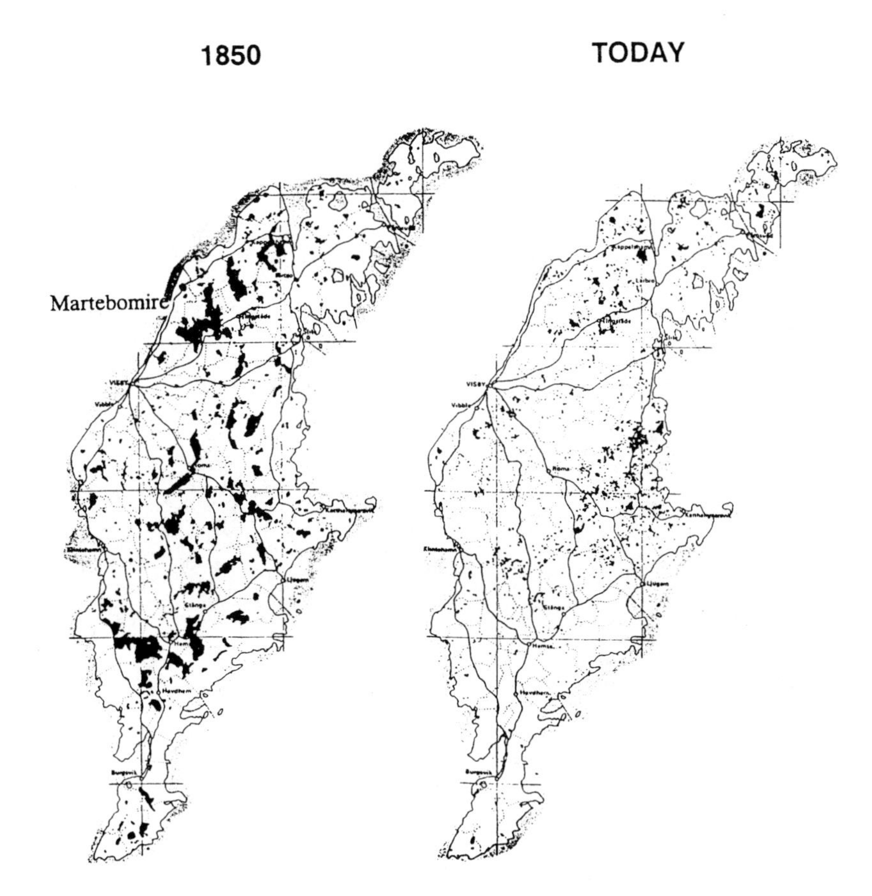

Figure 2. Wetland coverage (dark areas) on the island of Gotland in the Baltic Sea, prior to the extensive draining activites, and at present.

Prior to exploitation, the Martebomire, the wetland which is the main subject of this chapter, was a hydrological complex consisting of several mires and shallow lakes, covering a total area of about 34 km², with some 11 km² of open water. Several draining and ditching projects have reduced the wetland area to about 2.5 km² (Lundqvist et al., 1940; Nordberg, 1983). The fragments of the former lakes are mainly covered with a dense shrubby vegetation. The few remaining patches of the mire hold flora and fauna which are of conservation interest, but which at the same time are being considered for further conversion to agricultural land or dam constructions (Nordberg, 1983).

Figure 3. A typical wetland on the island of Gotland. From a photo by Gösta Håkansson.

2.2 Deterioration of the Wetland's Societal Support

In the ancient local farming the Martebomire wetland was harvested for hay and grazed by cattle and sheep, and wetland vegetation was also used as roof cover. The wetland lakes provided fish such as eel, pike, perch, burbot and various whitefish, and waterfowl (Jansson & Zucchetto, 1978; Folke, 1983). The Lummelunda marsh of the unexploited Martebomire had the largest surface area of any lake on the island of

Gotland. The mire complex served as a habitat for wetland-dependent flora and fauna. Among those, six species of orchids, seven other rare plants, and some ten bird species are today classified as endangered (Folke, 1983), with associated losses of genetic material and species diversity.

The peat (accumulated primary production) stored in the wetland, is an important part of the wetland system, and it contributes to the maintenance of environmental functions such as sustaining water quality and nutrient filtering. In the mid 1800's the homogeneous peat layer of the unexploited Martebomire was estimated to cover about 80 per cent of the wetland to an average depth of about 1.5 m (ranging from 0.5 to 6 m)(Lundqvist et al., 1940; Nilsson, 1982). The drained wetland provided very fertile arable land, due mainly to the high content of nutrients in the peat.[5] When the anaerobic peat environment of the wetland switched to aerobic conditions, decomposition of the peat layer started, and accelerated due to intensive cultivation and winderosion. The reduction of the peat layer and repeated floods have made it necessary to re-drain the wetland several times. In the 1980's, only about 10 km^2 of the peat area, and some 0.5 m of the peat depth remained (Nilsson,1982). The layer below the peat generally consists of rather poor soils (von Post, 1929). Increasing amounts of fertilizers have been used on the agricultural lands, partly to compensate for the decreased soil quality.[6]

The water supply in the area has decreased, especially during dry summers. This is due primarily to the reduced capacity of the drained and degraded wetland soils to store water.[7] Areas bordering the drained wetland have also been negatively affected (Petterson, 1946). There is no infiltration to the groundwater during dry periods. When people have drilled the wells deeper salt water intrusion has occurred[8] (Nilsson, 1982).

The severe water situation on the island as a whole has actualized community plans to build a pipeline from a distant lake to supply the largest urban area on the island with drinking water. If the wetland had not been exploited, part of the water from its lakes could have been used for this purpose.

There has been a gradual decrease in the quality of both surface and groundwater in the area. The increased use of artificial fertilizers and manure from domestic animals, and the leakage of nitrogen from the decomposing peat have resulted in nitrate pollution of the groundwater.[9] In addition, pesticides have been detected in the drinking water, and this has caused great alarm.

Today, the main drainage channel from the mire is connected directly to the riverbed of the associated Lummelunda stream. In addition to the fertilization from agricultural land the stream receives mechanically treated sewage. The phosphorus levels in the stream have only increased slightly, but there is a considerable increase in nitrogen, especially during the winter (Nilsson, 1982). This has caused a severe eutrophication of ditches and stream waters, and an export of nutrients and organic material to coastal waters.

The average water flow in the Lummelunda stream has decreased since the drainage took place. Today, the water flow fluctuates considerably with peaks of 3,000-6,000 l/sec during autumn, winter and spring, and low values of 10 l/sec during summer. The potential evaporation from the wetland, significant for the local precipitation, especially during dry summer months (Lundqvist et al., 1940), has been lost in the increased run-offs of water.

The stream produces about 2,500-3,000 anadromous trout (Salmo trutta) per year and holds a minor stock of rainbow trout (Oncorhynchus mykiss)(Westin et al., 1982). The eutrophication makes it necessary to clear the stream and stream banks regularly from vegetation. The fish stocks are threatened (with associated losses in commercial and sport fishing) by insufficient water flow and degraded water quality (c.f. Williams & Eddy, 1989). The conversion of the Martebomire has reduced the potential for sport fishing, bird watching, boating and other types of recreation in the area, and amenity values have also been lost.

The various exploitation effects that have occurred so far are summarized in Table 1.

2.3 Energy Analysis of Lost Wetland Support

The total amount of energy captured via photosynthesis is one approach to estimate the ecosystem's potential to do useful work for the economy (Odum, 1971; Turner et al., 1988; Costanza et al., 1989). Gross Primary Production (GPP) or alternatively, solar fixation is often used as an index of this captured energy.[3] The fixation of solar energy by plants and algae is necessary to maintain the wetland's functions and produce its environmental services (see Figure 1). I use the estimated GPP of the whole unexploited Martebomire, and compare it with the GPP of the present land use in the drained area, in order to obtain an approximation of the lost life-support value of the wetland.

Peat represents stored natural capital, derived from solar energy fixation, that has accumulated over thousands of years, and that contributes to the maintenance of the wetland's functions. I estimate the energy content of the stored natural peat capital prior to exploitation, and compare this estimate with the loss of peat energy. It has to be stressed that the energy content of the peat has no direct relation to the energy required to produce the peat.[2] GPP reflects the cost of production while the energy content of peat only reflects a minor part of the production costs.

2.4 Energy Analysis of Man-made Substitutes

The loss of the wetland's functions has created needs for replacement technologies. Of the technologies listed in Table 1, most of them already exist, some are planned to take place in the near future, or they exist in similar drained areas on the island. A few technologies have been used to illustrate "shadow" replacements of the lost wetland services.

Monetary Replacement Costs

Information about the monetary costs of man-made substitutes was obtained from landowners, governmental agencies, private firms and from various statistical sources (see notes to Table 3). The monetary estimates, used as the basis for the energy analyses, are simply a straightforward undiscounted summation of the costs associated with replacing the wetland's functions, and they are expressed in 1989 Swedish Crowns (SEK[10]). The actual depreciation of capital costs[11] varies from 10 to 50 years.

Energy Analysis Approaches

Energy is necessary for all production activities, and impossible to create internally or recycle. It must be supplied from outside the system, whether economic or ecological. All economic production is associated with a physical cost. This cost can be estimated in energy terms by tracing the flows of physical materials in the economy.

I make two separate analyses of the embodied industrial energy cost[2] that is required to replace the wetland's societal support (description of energy analysis methodology can be found in e.g. Bullard et al. 1978; Costanza & Herendeen, 1984: Hannon et al., 1985). Ideally, energy intensities[12] of the Swedish economy would have been used. At present, energy intensities are available only for a 1980, 24-sector input-output model[13] of the Swedish economy (Östblom, 1986), which is too aggregated for the purpose of this chapter.

The most detailed energy input-output tables are from the U.S. economy where, in the first approach, energy costs were derived from the 1977 energy input-output table for 398 sectors of the U.S. economy (Hannon et al., 1985). This table uses the same system boundaries as the input-output tables developed by Leontief (1966) for analysing the costs of producing goods and services in the various sectors of the economy. This approach has consumption expenditures, government services and net exports as outputs as well as investment (Cleveland, 1988).

In the second analysis, energy intensities were based on the extended energy input-output model of 1972, covering 87 sectors of the U.S. economy (Costanza & Herendeen, 1984). This extended input-output table includes energy costs of government and households in the input-output model, which according to Cleveland (1988), implies that the only major output of such an economy is gross capital formation.

The energy input-output tables use coal, crude oil and gas, refined petroleum, electricity, natural gas (Hannon et al., 1985) and industrial, hydro- and electric energy (Costanza & Herendeen, 1984) as energy inputs to the economy, and correct them to include the fuel needed to extract and process the primary energy source. The energy use per dollar derived from the input-output tables has been corrected for inflation and currency conversion, and is expressed in 1989 SEK.

Table 1. The wetland Martebomire's support to society, exploitation effects, and replacement technologies.

Societal Support	Exploitation Effects	Replacement Technologies
Accumulating and storing organic matter (peat)	Peat layer reduction and disapperance through decomposition, intensive farming, and wind erosion Degraded soil quality	Artificial fertilizers Re-draining of ditches
Maintaining drinking water quantity	Reduced water storage Lost source for urban area	Water transport[a] Pipeline to distant source[b]
Maintaining drinking water quality	Nitrates in drinking water Pesticides in drinking water	Water-quality inspections Water purification plant Silos for manure from domestic animals Nitrogen filtering[b] Water transport
Maintaining ground water level	Dried wells Saltwater intrusion	Well-drilling Saltwater filtering[b]

Maintaining surface water level	Decreased evaporation and precipitation, reduced amount of water	Dams for irrigation Pumping water to dam Irrigation pipes and machines Water transport for domestic animals[a]
Moderation of waterflows	Pulsed run-offs Decreased average water flow in associated stream	Regulating gate Pumping water to stream
Processing sewage, cleansing nutrients and chemicals	Reduced capacity Eutrophication of ditches and stream	Mechanical sewage treatment Sewage transports Sewage treatment plant Clear-cutting of ditches and stream
Filter to coastal waters	Adding to eutrophication Adding to silting up	Nitrogen reduction in sewage treatment plants[b]
Providing		
- food for humans	Loss of food sources	Agriculture production Imports of food[c]
- food for domestic animals	Loss of food sources	Agriculture production Imports of food[c]
- roof cover	Loss of construction materials	Roofing materials[c]

(Table 1 continues on next page)

Sustaining:		
- anadromous trout populations	Degraded habitat, commercial and sport fishery losses	Releases of hatchery raised trout[d] Farmed salmon[d]
- other fish species	Loss of habitat	Work by non-profit organizations
- wetland dependent flora and fauna	Loss of habitat, Endangered species	
Species diversity	Losses	
Storehouse for genetic material	Losses	
Bird watching, sport fishing, boating and other recreational values	Losses	
Aesthetic and spiritual values	Losses	

Notes to Table 1

a. these replacements are common in similar drained areas on the island of Gotland.
b. these replacements are being planned; they are further discussed in relation to Table 3.
c. not further considered in the analysis because of difficulties in quantifying the losses of food sources and construction materials.
d. there is some natural reproduction in the stream, and also occasional releases of hatchery raised trout; the cost of farming salmon is a measure of the commercial losses, due to the reductions in the amounts of returning adult trout.

2.5 Solar and Industrial Energy Expressed in Terms of the Same Quality

Solar and industrial energies are not directly comparable. Plant biomass is a less concentrated form of energy than coal or oil. Peat is a fossil fuel that can be used for electric power production with an efficiency here assumed to be similar for coal. The fixed solar energy (GPP) must be converted to fossil fuel equivalents (FFE) to make comparison possible between the life-support evaluation of the wetland and the cost for the man-made substitutes. To transform solar energy based production to fossil fuel energy equivalents is a controversial step. Lavine (1984) concluded, from examiningeight studies relating coal and embodied sunlight energy values, that the appropriate range is 71-700 calories of sunlight per calorie of coal, derived from a study of the major uses of coal and sunlight in ecological and economic systems, i.e. to develop and maintain a low entropy gradient in a water cycle. However, the present analysis requires an energy quality comparison between GPP and fossil fuels. Studies comparing electricity production in a steam-powered electric plant via coal-burning in comparison with sunlight embodied in wood-burning and in comparison with sunlight used in a blue-green algae mat producing electricity were investigated in Lavine's paper (1984). The comparisons have been questioned because they assumed that electricity production is an equally effective use of both coal and wood, or coal and algae. In a revised version of Odum & Odum (1981) the transformity (the ratio between the amount of energy of one type required to generate energy of another type) from GPP to fossil fuels is estimated to be 10-13, derived from relationships between solar insolation and the total workings of the biosphere. I use these globally based transformities[14] as an approximation for expressing GPP in terms of fossil fuel equivalents. It must be emphasized that all energy transformation ratios expressing industrial energy flows as solar energy or vice versa, must be considered first-cut estimates pending more exhaustive tabulations of transformations from sunlight to coal or other fuels (Lavine & Butler, 1981).

3. RESULTS

The annual direct plus indirect industrial energy cost for the technical replacements is estimated to be 15-35 and 20-50 TJ FFE, which thus approaches the loss of solar fixing ability of the exploited wetland which is estimated to be 55-75 TJ FFE per year. I estimate the average annual loss of the energy content of the peat, over the 130 years since exploitation started, to be as much as 7,985 TJ FFE (Table 4). The undiscounted annual monetary cost of the replacements is about 2.5 to 7 million SEK (US$0.4 to 1.1 million)(Table 3), or about 2,500 to 5,500 SEK annually per person living in the affected area.[15]

Table 2. Annual gross primary production (GPP)[a] in the wetland Martebomire prior to exploitation in the middle of the 19th century (unexploited wetland), and at present (exploited wetland). MJ (10^6J), TJ (10^{12}J).

	km^2	GPP· m^{-2} yr^{-1} MJ	Total GPP yr^{-1} TJ
Unexploited wetland			
mires	23	69.0	1,587
lakes	11	12.6	139
Total	**34**		**1,726**
Exploited wetland[b]			
arable land[c]	22	29.6	651
grazed and pasture land	2	29.0	58
coniferous forest[d]	5	30.0	150
mixed deciduous forest[d]	1	38.0	38
mire vegetation[e]	1.2	69.0	83
lake fragments[e]	1.2	12.6	15
non-productive land[d]	1.6	-	-
Total	**34**		**995**
Total annual loss of GPP			**731**

Notes to Table 2:

a. from Jansson and Zucchetto (1978).

b. about 70 per cent (25 km^2) of the former Martebomire was located within the Lummelunda drainage basin (60 km^2). There are detailed land use data from the drainage basin (Nordberg, 1983), on which the present Martebomire land use analysis is based.

c. estimating that 65 per cent of the total Martebomire has been converted to arable land, which is also the average percentage for drained wetlands on Gotland.

d. based on Norberg (1983, Table 3). Non-productive land; infra-structure, populated and industrial areas, abandoned cultivated land etc.

e. based on Lundqvist et al. (1940), Nilsson (1982), Nordberg (1983).

3.1 Gross Primary Production and Peat Losses

As shown in Table 2, the unexploited wetland consisted of mires and lakes that have been transformed mainly into arable land but to some extent also into forests. I estimate the total annual GPP prior to the wetland's transformation to be 1,726 TJ, and the present GPP to 995 TJ, indicating a reduced solar fixing ability of 730 TJ per year, which is a loss of almost 45 per cent of the original GPP.

The energy content of the wetland's peat layer prior to exploitation is estimated to have been 1,184.1 PJ, and the energy content of the remaining peat to have been 146.2 PJ, or almost a 90 per cent loss of the peat energy. Assuming that a similar amount of peat has deteriorated each year over the past 130 years, through decomposition, intensive agriculture and wind erosion, this gives us an annual loss of stored natural capital of 7,985 TJ (see note c in Table 4).

In Table 4, the annual loss of the wetland's life-support and peat energy is compared with the industrial energy replacement cost, all values expressed in FFE.

3.2 Industrial Energy Replacement Costs

Technologies aimed at replacing the wetland's biogeochemical processes dominate the monetary cost (82-52 per cent of total costs) estimated to be 2.5 to 7 million SEK per year (Table 3), followed by replacements for the hydrological cycle (7-40 per cent of total costs), and the biological system (8-11 per cent of total costs). Artificial fertilizers are the single most costly replacement. Based on empirical data on decreasing grain yields per kg of fertilizer use (Zucchetto & Jansson, 1985) the cost for increased use of artificial fertilizers here, is assumed to be attributed to the deterioration and erosion of the peat layer (Table 3, note i). Water transport, sewage transport and treatment costs are based on the approximately 1,000 persons, and the actual number of domestic animals living in the affected area (for futher details see the notes to Table 3).

Between 71-91 per cent of the annual replacement cost of 17-37 TJ, derived from the 1977 U.S. energy input-output table, consists of losses associated with biogeochemical processes, to 4-25 per cent of losses associated with the hydrological cycle, and to 4-5 per cent of losses of the wetland's biological system.

The annual energy cost of replacements based on the 1972 U.S. energy input-output table is estimated to be 18-48 TJ. Of these energy costs, 54-84 per cent aim at replacing the biogeochemical processes, 6-38 per cent the hydrological cycle, and 8-10 per cent the wetland's biological system.

The percentages for replacements of the biogeochemical processes, the hydrological cycle, and the wetland's biological system from the 1972 U.S. energy input-output table, with the extended system boundary that includes the household and governmental sectors, are in accordance with the monetary percentages. This is not the

Table 3. Monetary and energy costs of wetland replacement technologies[a], expressed in thousands of SEK[b] 1989 and in fossil fuel equivalents TJ (10^{12}J).

Replacement Technologies[c]	Monetary Costs[d]	Energy Costs	
		1977 Input-Output Analysis[e]	1972 Extended Input-Output Analysis[f]
	1,000 SEK	TJ	TJ
Re-draining and clear-cutting[g] of ditches and stream (G)	50- 56	0.200- 0.220	0.295- 0.330
Dams for irrigation[h]	57-205	0.200- 0.720	0.325- 1.167
Pumping water to dams	11- 36	0.025- 0.080	0.070- 0.225
Irrigation pipes and machines (H)	58-184	0.230- 0.735	0.330- 1.045
Artificial fertilizers (G)[i]	880-1428	12.495-20.275	7.650- 12.415[u]
Regulating gate[j]	10- 18	0.060- 0.105	0.110- 0.200
Pumping water to stream (H)	7- 11	0.015- 0.025	0.045- 0.070
Well-drilling (H)[k]	33- 53	0.080- 0.130	0.190- 0.300
Water-quality inspection[l]	12- 40	0.020- 0.070	0.055- 0.180
Water purification plant (G)[m]	32	0.230	0.180

Nitrogen and saltwater filtering (H)[m]	0 -460	0 - 3.290	0 - 2.570
Water transport (H)[n]			
- for humans	0 -500	0 - 1.295	0 - 4.050
- for domestic animals	0 -335	0 - 0.870	0 - 2.715
Pipeline to distant source (H)[o]	0 -990	0 - 1.955	0 - 5.570
Mechanical sewage treatment and storage (G)[p]	625-750	1.360- 1.630	3.555- 4.265
Silos for manure from domestic animals (G)[q]	370-950	1.160- 2.975	2.105- 5.405
Sewage transport[r]	63-126	0.165- 0.325	0.510- 1.020
Sewage treatment plant	3-200	0.005- 0.365	0.015- 1.170
Nitrogen reduction in sewage treatment plants (G)	40- 45	0.085- 0.100	0.230- 0.255
Hatchery-raised trout and farmed salmon (B)[s]	210-250	0.720- 0.855	1.400- 1.670
Endangered species (B)[t]	68-314	0.120- 0.550	0.425- 1.965
TOTAL	**2530-6985**	**17-37**	**18-48**

Notes to Table 3.

a. see Table 1 concerning the wetland's societal support, exploitation effects, and replacement technologies.

b. in 1989 one US dollar was about 6.43 Swedish crowns (SEK).

c. technologies attributed to losses in the performance of biogeochemical processes (G), the hydrological cycle (H), and to the biological part (B) of the wetland ecosystem; there are no sharp boundaries as these performance values are interdependent (see Figure 1).

d. based on actual data from landowners, private firms, government agencies, and other institutions,.

e. based on energy intensities from Hannon et al. (1985).

f. based on energy intensities from Costanza & Herendeen (1984); energy intensities from sectors similar to the 1977 input-output data have been used as far as possible. In the other cases, an average energy intensity for all sectors except energy, fisheries, agriculture, and forestry is used (Costanza, pers.comm.).

g. 24.2 km ditches drain an area of 71 km^2; the peat area covered about 28 km^2; assuming 50 per cent of the ditches are in the downstream section of this area, with one fourth blown; re-draining and clear-cutting through machinery every five to ten years.

h. one dam was built in 1974, used for the irrigation of 3 km^2; two more are planned assuming similar costs for pumps and irrigation; the lower value relates to the present dam, the higher value to three dams.

i. between 1900 and 1980 arable land increased 1.3 times on Gotland, and fertilizer use 13 times; between 1950 and 1980 the yield of grains decreased by 1.8 kg per kg fertilizer use (Zucchetto & Jansson, 1985, Table 2.1); the decrease is assumed to be mainly the result of soil deterioration and erosion of the peat layer and associated water losses; the increased use of fertilizers per grain yield (about 200 kg per ha) is used here as an approximation of efforts to replace land degradation; the estimate is based on the application of fertilizers on various croplands on Gotland (on average annually 500-600 kg per ha with an average yield of grain of about 2 tonnes per ha); arable land use in the study area is similar to arable land use in Gotland as a whole (Nilsson, 1982; Jansson & Titus, 1984).

j. 2.5 x 5 m of iron, three pumps with a capacity of about 320 l/min.

k. includes electric pumps and other equipment; average depth for wells on the island is 35-40 m, in wetland areas about 20 m; 50-60 new wells have been drilled since 1976, there were 15-30 wells in the area prior to 1976; the higher value includes the 15-30 wells assuming that they have been re-drilled.

l. inspection of agricultural chemicals, nutrients, bacteria; today about 4 per cent of the wells are inspected, the higher value assumes inspection of 25 per cent.m. small water purification plant in Martebo serving 50 persons; there have been high levels of nitrate in groundwater, and saltwater intrusion has occurred in this area; assuming 25 per cent of the wells (see note k) and water consumption will be affected; estimates concern nitrogen filtering or nitrogen plus saltwater filtering of wells (Folke, 1983); saltwater filtering is expensive and nitrogen filtering alone accounts for only 10-20 per cent of the estimate.

n. although common on the island there has been no recorded water transport to the area so far; last summer the average groundwater level was lowered by 30 m; assuming further degradation of the peat will cause water deficiency for a period of two weeks in dry summers; assuming that the water is transported 30-40 km; average water consumption for humans ranges from 150-300 l/24h, we use 200 l/24h; about 1000 persons are affected; water consumption by domestic animals (cows, calves, pigs, sheep, and chicken) from (Hilding, 1982).

o. due to problems with drinking water support to Gotland's capital there are plans to build a pipeline to a distant lake, one of the few remaining lakes - the Bästeträsk- on the north-eastern part of the island; prior to exploitation the largest lake by area was situated within the Martebomire, and there were additional smaller lakes; it is hard to tell whether the additional water needs of the capital could have been supplied by these lakes without causing many of the environmental effects discussed in the paper; the higher cost assumes that this would have been possible, represented by a reduction by 15 per cent assumed to be the cost for a pipeline to the Martebomire.

p. four persons per household and two alternative sewage treatment technologies; about 1000 persons in the wetland area.

q. about 20,000 m^3 manure from domestic animals and silos of 200 m^3 to 1000 m^3.

r. 3-4m^3/tank cleared 1-2 times a year; 4 persons per tank; about 1000 persons from the wetland area; the sewage is transported 30-40 km; treatment plant: lower value clearing of sewage per m^3, higher value average costs from plants with a capacity of 25,000-125,000 person equivalents (pe); Visby plant 47000 pe, 90 per cent reduction of phosphorous, 35 per cent of nitrogen; there will be 50 per cent nitrogen reduction in the Visby plant.

s. 2,500-3,000 smolts annually produced in the stream; there have been occasional releases of hatchery raised trout; the profit from harvesting returning trout in the stream was 12,000-30,000 SEK in 1955 corresponding to a value of 5-12 tons of farmed salmon in 1989; average recapture is about 15 per cent or 600 kg per 1,000 salmon smolts released (Folke, 1988), which is about 1-2 tons; the value of this catch is substracted from the 1955 value; present sport fishing values are not considered.

t. costs in non-profit making associations for trying to save endangered birds, orchids and other plants that have disappeared from or presumably will disappear from the area (Folke, 1983); it is doubtful whether these species would have ceased to be endangered if the wetland had not been exploited; I assume that 5-10 similar wetland habitats are required.

u. artificial fertilizers as sector 173 (179710 BTU/$) in Hannon et al. (1985), but included in the chemical product sector (148830 BTU/$) in Costanza & Herendeen (1984).

case for the 1977 U.S. energy input-output table, based on traditional economic system boundaries. In the latter there is one sector specified for fertilizers, which is not the case in the 1972 U.S. energy input-output table, and may be one reason for the discrepancy (see note u, Table 3). Another reason is that households and government sectors are included in the 1972 table (Costanza & Herendeen, 1984).

3.3 Comparison of Life-Support Losses and Man-Made Substitutes

Using the revised transformities of 10 to 13 GPP/FFE (Odum & Odum, 1981) gives an estimate of 55-75 TJ FFE for the annual GPP loss.

Comparing the annual loss of GPP with the annual cost for the replacement technologies (Table 4) indicates that the industrial cost for the substitutes approaches the loss of the environmental functions previously sustained by the GPP. They are within the same order of magnitude, with the energy loss of GPP-based life-support being 1.5-3.5 times larger than the energy use for producing and maintaining the replacements.

As stated above, the major quantitative energy loss is associated with the diminished natural peat capital. I estimate that the annual loss of peat energy is two orders of magnitude larger than the energy cost for the replacement technologies (Table 4).

4. DISCUSSION

The results show that it takes considerable amounts of money and costly industrial energy to replace the loss of environmental functions. These environmental services were previously provided without cost by the wetland system. The major part of the costs, whether monetary or biophysical, is calculated as technical substitutes for the biogeochemical processes in the wetland, followed by substitutes for the services associated with the hydrological cycle. Not more than about 10 per cent are directly related to the biological part of the system, which perhaps illustrates a lower priority and difficulty of substituting for the losses of species diversity and genetic variability, losses that are generally irreversible.

The annual cost of the industrial energy-dependent technologies is approaching the annual loss of the solar energy fixing ability of the wetland. Despite this, *the replacements do not substitute for the wetland's life-support*, since there are still severe environmental problems in the area. In addition, almost 90 per cent of the peat, the GPP accumulated over thousands of years, has been destroyed, corresponding to an annual estimated energy content value 160-530 times larger than the annual industrial energy costs. This large difference, which would have been even greater if the cost of producing the peat would have been included in the estimate, is due to the fact that the peat layer has diminished over a short period (130 years) relative to its creation (1,500 years). When the peat is eliminated it is gone forever, but the replacement technologies will remain, and, thus, the differences between peat energy losses and industrial energy costs will decrease and eventually shift with time. As discussed in sections 1.1 and 2.2 the peat is a crucial component for maintaining valuable wetland functions.

The many wetland functions discussed in this paper seldom have a direct market value. This is one explanation for the fact that the wetland's often unperceived but real and long-lasting societal support values have been consumed or converted to land use activities that generate a short-term, direct capturable and immediate income stream. The technical replacements are made because of a need in society, but at this stage it is seldom perceived that the need has appeared due to the loss of the wetland's life-support. Furthermore, the industrial energy based techologies only partly replace the wetland's functions, and they are only point solutions, not integrated ecological economic solutions.

The values of the environment as a factor of production, and the free support of ecosystems are not yet fully recognized within economic systems, although economies are dependent on this support in order to be able to function (Daly, 1984; Odum, 1989). This is one reason why it is difficult to capture the significance of the life-support through human based preferences valuation alone. Willingness to pay is not coupled directly to the fundamental role of ecosystems in sustaining socio-economic activities. Therefore, a monetary estimate, discounted or not, does not indicate whether we have covered one per cent or ninety-nine per cent of the actual life-support value. There is a need for supplementary estimates of the biophysical foundations and interrelations between ecosystems and economic systems.

This is clearly illustrated by the results. The estimate of the monetary replacement costs tells us that the annual (undiscounted) replacement cost for the wetland's functions is about 2.5 - 7 million SEK (US$0.4 - 1.1 million). If we are to compare this estimate with the benefits of alternative uses of the wetland, we would still have no idea of the magnitude of the "true" societal value of the wetland's life-support. This is even more so if these alternatives are based on unsustainable options (Daly & Cobb, 1989), such as the present intensive type of agriculture, and also if the technical substitutes do not fully replace the multifunctionality of the wetland system, as is the case in this particular study.

The energy analysis applied in this chapter aims at indicating the size of the actual life-support. I do this by comparing the energy used throughout the economy to produce and maintain the replacement technologies by the energy required by the wetland system to produce and maintain similar environmental functions.

There are, of course, a lot of uncertainties in an estimate like this. First of all the wetland GPP is very approximate, and it assumes a fairly constant annual solar fixing ability over the past 130 years. Furthermore, one might question the usefulness of GPP as an estimate of the wetland's life-support value. It shows the work contribution to the human economy of the living part of the wetland system. This work is required to generate and maintain many of the wetland's environmental functions, also those associated with biogeochemical processes and the hydrological cycle. Without this work the non-living parts of the system would not be modified to sustain the environmental functions (Figure 1). However, just as human systems are dependent on non-living environmental services, such as water and its associated minerals, this is also true for the wetland system. The solar energy embodied in such flows is considerable (Odum & Odum, 1981). Furthermore, the GPP method does not take into account the interdependence between ecosystems or differences in productivity within the same habitat. Also, an ecosystem with relatively lower GPP may be an important node in the regional landscape, for example as a water supply system for adjacent ecosystems or developed land (Jansson, 1988; Turner et al., 1988).

Table 4. Annual loss of wetland productivity and stored natural capital, the peat, and the annual energy cost of technical replacements, TJ (10^{12}J).

	FOSSIL FUEL EQUIVALENTS[a] **TJ per year**
ECOSYSTEM LOSSES	
Loss of Gross Primary Production[b]	55 - 75
Loss of Peat[c]	7,985
ECONOMY REPLACEMENTS	
Costs of Technical Substitutes[d]	15 - 35
	20 - 50

Notes to Table 4:

a. see sections 2.5 and 3.3.

b. see Table 2.

c. energy content of Gotland peat 23 kJ per g; total wetland area on Gotland prior to exploitation 236 km^2, and energy content of 1 m peat of these wetlands 6,900 PJ (Zucchetto & Jansson, 1985); stored peat capital 1.5 m depth over about 27 km^2 (Lundqvist et al., 1940; Nilsson, 1982). Annual peat accumulation about 1mm (Jansson & Zucchetto, 1978), assumed to be included; energy content unexploited wetland 1,184.11 PJ (10^{15}J); remaining stored capital 0.5 m depth over about 10 km2 (based on Nilsson, 1982); energy content exploited wetland 146.19 PJ; annual loss of stored peat capital 1,184.11 PJ - 146.19 PJ divided by 130 years (about 1850-1980).

d. see Table 3.

The industrial energy estimates are very approximate, since by using the U.S. input-output tables I implicitly assume that there are structural similarities between the U.S. and Swedish economies, and that no increase in the efficiency in total energy use (direct plus indirect) has taken place since the years of the tables. I believe, however, that these methodological deficiencies are acceptable for the purpose of this chapter.

Furthermore, some of the replacements are related to institutional rules and structural changes in the economy, and it is hard to tell whether they would have taken place independently of the loss of the wetland's functions. Nevertheless, they illustrate environmental services that could have been provided completely without monetary and fossil energy costs by wetlands. More important, *the current replacements do not fully cover the loss of the wetland's functions*. For example, the role of cleansing water of chemicals and pollutants, the accumulating problems of sludge and ash from sewage treatment plants, and the eutrophication of the wetland drainage basin have not been accounted for fully, which is also the case for the severe losses of species richness and genetic diversity. Furthermore, recreational, aesthetic and spiritual values are not covered by the energy estimate. In principle, it would be possible to include recreational values based on the travel-cost method, since travelling requires energy for the cars, boats, aircrafts etc. It is of course impossible to estimate an energy cost for monetary estimates derived from artificial markets (see Bojö, Chapter 3 of this book). Such estimates are not connected directly to economic activities that actually take place in society and that thus have no associated true energy cost. Option, existence, and bequest values (e.g. Nash & Bowers, 1988; Pearce et al., 1989) are examples of estimates of this type, that attempt to cover the monetary value of natural environments. Although relevant, such valuation methods are also based on subjective human preferences which, as previously stressed, are not directly coupled to the socio-economic dependence on the life-supporting ecosystems (see Folke, Chapter 5 of this book).

It is often believed that the negative effects of a project tend to diminish with time. The present wetland study has illustrated that this is not so. The environmental effects of the wetland exploitation tended to accumulate and show up several decades or even more than a century later. A cost-benefit assessment back in the 1850's would most surely have underestimated the life-support values, because they were unknown, and the environmental goods and services were taken for granted. We are presumably in a similar situation today when we try to evaluate the environment. According to Pearce et al. (1989) "a social objective based on mankind's more immediate well-being need not be consistent with long-run welfare or even human survival...Some care needs to be exercised, then, that the use of social objectives such as gain in welfare does not dictate or support policies which are inconsistent with the ecological preconditions for existence...". One should thus be very wary of undertaking projects with irreversible effects, such as the destruction of ecosystems and loss of species, "even when a conventional cost-benefit analysis would indicate that it is desirable to proceed with the project. In these circumstances, it would seem appropriate to require that cost-benefit analysis be used to choose between alternatives only within a choice set bounded by some sensibly chosen environmental constraints" (Nash & Bowers, 1988). Although it has a lot of weaknesses the present analysis is an attempt in that direction. Although the accuracy of the estimates could be considerably improved, we get an idea of the biophysical interrelations between a wetland system and the surrounding economy, and the value of life-support. Further methodological developments, energy or non-energy based, are needed to provide insights and more accurate measures of the dynamic interdependencies of ecological and socio-economic systems.

There is also an urgent need for evaluations that do not only consider the environment when the environmental damage has already been done. The method described in this paper, and the ones applied by Gosselink et al. (1974) and Costanza et al. (1989), make it possible to assert and include in analysis the life-support of ecosystems before it is destroyed. This support is necessary for a prosperous economy and a sustainable development (Daly, 1984), and it includes many essential functions for societal welfare, independent of whether human preferences recognize the environmental contributions or not. Technical solutions can only to some extent replace the ecosystem support, and at a high cost since expensive non-renewable fossil fuels are required for the replacement (Odum, 1979). From a societal perspective it would be much wiser to use these "replacement fuels" to maintain and enhance the life-support of ecosystems, instead of trying to replace it when it has been destroyed. One solution in this direction would be to apply ecological engineering (Odum, 1971; Mitsch & Jörgensen, 1989) to reconstruct parts of the wetland, and use it for drinking water supply, aquaculture, recreation, as a processor of sewage sludge, or as a filter for nitrogen to coastal waters (Nichols, 1983; Ewel & Odum, 1984; Fleischer & Stibe, 1989; Jingsong & Honglu, 1989; Mitsch et al., 1989; Shijun & Jingsong, 1989; Andréasson-Gren, Chapter 9 of this book). Ecological engineering explicitly makes use of the self-designing capability of ecosystems, counteracts environmental degradation and makes it possible to redirect expensive and valuable industrial energy, materials and processes to other parts of society.

5. CONCLUSION

Ecosystems are often unacknowledged subsidies to the economy, but their life-support is of high value to society. There is an urgent need to find evaluation methods that make it possible to consider this support. A monetary valuation does not reveal whether I have estimated one thousandth or 99 per cent of the actual life-support. The energy analysis approach applied in this chapter provides a complementary biophysical estimate. Although there is a need for improving estimates and methods this type of analysis indicates and makes it possible to compare the relative contribution from ecosystems and economies for producing and maintaining the necessary goods and services. In general, industrial energy based technologies only partly replace the environmental functions, and are only point solutions, not integrated ecological economic solutions to environmental problems. Instead of having to replace the environmental functions when they have been destroyed, industrial energy based technologies should be developed to supplement and enhance the ecosystem life-support, thereby reducing and reallocating the use of costly and often scarce industrial energy and natural resources to other parts of society.

ACKNOWLEDGEMENTS

This study would not have been possible without the huge amount of empirical data and analyses performed in the Gotland project led by Ann Mari Jansson. Ann Mari has contributed a great deal to this paper, not only due to her extensive knowledge of ecology and economy interrelations and of the island of Gotland, but also through her enthusiasm and interest in my work. Many thanks are also due to Nils Kautsky, Ing-Marie Andréasson-Gren, and Folke Günther, and to the referees Cutler Cleveland, Charles Hall and Robert Herendeen for most valuable comments on the manuscript. An earlier version of this chapter was a part of Folke, (1990). Funding was provided from the Swedish Council for Forestry and Agricultural Research.

NOTES

1. energy is essential in the production of all goods and services, whether natural or man-made. It is impossible to produce anything without energy.
2. embodied industrial energy refers to the direct (e.g. fuel energy, electricity at the site) and indirect (energy used in previous upgrading stages) energy required to produce economic goods and services. The embodied industrial energy is the amount of direct and indirect uses of oil, coal, gas, hydro- and electric energy, expressed in fossil fuel equivalents, e.g. it takes 3-4 units of fossil energy to generate one unit of electricity. In this chapter the embodied solar energy is estimated as the solar energy fixed by plants and algae, required for the work of the food-web. In addition to the direct solar fixation there are also other solar derived energies such as freshwater flows, winds, etc. which maintain the structure of the wetland and affect the GPP. These energies are not estimated here, but they are of high significance for sustaining both ecological and economic systems (Odum & Odum, 1981).
3. it should also be stressed that the wetland system is an integrated part of the landscape, and its production of environmental services is thus dependent on what is taking place in adjascent ecosystems.
4. peat-accumulating wetlands that receive some drainage from surrounding mineral soil and usually support marshlike vegetation (Mitsch & Gosselink, 1986).
5. about 150-200 g N m^2 in the upper cm at the beginning of 1900 (Nilsson, 1982). Peat has also been cut in the area, but not in any large scale.

6. the present grain yield on the island of about 2 tonnes/ha decreased between 1950 and 1980 with 1.8 kg/kg fertilizer used (Zucchetto & Jansson, 1985, Table 2.1) (see Table 3, note i).

7. the ability of the area to store water decreased by 25 mm from 1931 to 1981 (Nilsson, 1982), indicating a total storage loss of at least 200,000 m^3 of water during this period (Folke, 1983).

8. high concentrations of Na and Cl have been found in well water in the area (Nilsson, 1982). In 1969 there had been saltwater intrusion in about 30 per cent of the Gotland wells (Folke, 1983).

9. the level of nitrate in the drinking water has ranged from undetectable to 100 mg/l, by far exceeding the acceptable 30 mg/l for healthy drinking-water according to WHO standards. Leakage is especially large during heavy rain and storms (Spiller et al., 1981; Nilsson, 1982; Jansson & Titus, 1984; Andréasson, 1989). Jansson & Titus (1984) indicated, via a simulation model of the drained Martebomire area, that there is a close relationship between the agricultural nitrogen put on moraine or sand and the nitrogen reaching the groundwater. This relationship was not equally clear for drained peat soils.

10. in 1989 one US$ was 6.43 Swedish crowns (SEK).

11. many of the capital costs are historical, but they represent the costs of replacing environmental damage and they are therefore included in the estimate.

12. energy intensity is the direct and indirect energy requirement to produce a unit of commodity.

13. input-output models are generally static and assume constant coefficients, average value equals marginal value, linearity etc. (see Costanza & Herendeen (1984) and Hannon et al. (1985) and references therein).

14. Table of Transformities (from Odum and Odum, 1981, revised by H.T. Odum, August 1987).

Type of energy	Solar transformity solar emjoules per joule	Coal transformity coal emjoules per joule
Sunlight	1	0.000025
Gross plant production	4,000	0.1
Coal delivered for use	40,000	1.0
Oil delivered for use	53,000	1.3
Electricity	160,000	4.0

15. all monetary costs except for the water pipeline, and endangered species that also concern others than the approximately 1,000 persons in the affected area.

REFERENCES

Andréasson, I.-M. 1989. Costs of Controls on Farmers' Use of Nitrogen: A Study Applied to Gotland. Dissertation for the doctor's degree in economics. The Economic Research Institute, Stockholm School of Economics, Stockholm.

Batie, S.S. and Shabman, L.A.1982. Estimating the Economic Value of Wetlands: Principles, Methods, and Limitations. *Coastal Zone Management Journal* 10:255-278.

Bullard, C.W., Penner, P.S. and Pilati, D.A. 1978. Net Energy Analysis: Handbook for Combining Process and Input-Output Analysis. *Resources and Energy* 1:267-313.

Cleveland, C.J. 1988. Physical and Economic Models of Natural Resource Scarcity: Theory and Application to Petroleum Development and Production in the Lower 48 United States. Ph.D. Dissertation. University of Illinois, Urbana, USA.

Costanza, R. and Herendeen, R.A. 1984. Embodied Energy and Economic Value in the United States Economy: 1963, 1967 and 1982. *Resources and Energy* 6:129-163.

Costanza, R., Faber, S.C. and Maxwell, J. 1989. Valuation and Management of Wetland Ecosystems. *Ecological Economics* 1:335-361.

Daly, H.E. 1984. Alternative Strategies for Integrating Economics and Ecology. In: Jansson, A.M. (ed.). Integration of Economy and Ecology: An Outlook for the Eighties. Proceedings from the Wallenberg Symposia. Askö Laboratory, Stockholm University, Stockholm. pp. 19-29.

Daly, H.E. and Cobb, J.B. 1989. For the Common Good: Redirecting the Economy Towards Community the Environment, and a Sustainable Future. Beacon Press, Boston.

Day, J.W., Hall, C.A.S., Kemp, W.M. and Yanez-Arancibia, A. 1989. Estuarine Ecology. John Wiley and Sons, New York.

Ewel, K.C. and Odum, H.T. 1984. (eds.) Cypress Swamps. University Presses of Florida, Gainesville.

Faber, S.C. 1987. The Value of Coastal Wetlands for Protection of Property against Hurricane Wind Damage. *Journal of Environmental Economics and Management* 14:143-151.

Fleischer, S. and Stibe, L. 1989. Agriculture Kills Marine Fish in the 1980s: Who is Responsible for Fish Kills in the Year 2000? *Ambio* 18:347-350.

Folke, C. 1983. Ekonomisk Värdering av Miljöförändringar: i Samband med Myrexploatering. Askö Laboratory, Stockholm University, Stockholm.

Folke, C. 1988. Energy Economy of Salmon Aquaculture in the Baltic Sea. *Environmental Management* 12:525-537.

Folke, C. 1990. Evaluation of Ecosystem Life-Support in Relation to Salmon and Wetland Exploitation. Ph.D. Dissertation in Ecological Economics. Department of Systems Ecology, Stockholm University, Stockholm.

Gosselink, J.G. and Turner, R.E. 1978. The Role of Hydrology in Freshwater Wetland Ecosystems. In: Good, R.E., Whigham, D.F. and Simpson, R.L. (eds.). Freshwater Wetlands: Ecological Processes and Management Potential. Academic Press, New York. pp. 63-78.

Gosselink, J.G., Odum, E.P. and Pope, R.M. 1974. The Value of a Tidal Marsh. Publication No. LSU-SG-74-03. Center for Wetland Resources, Louisiana State University, Baton Rouge.

Hammack, J. and Brown, G.M. 1974. Waterfowl and Wetlands: Towards Bioeconomic Analysis. The Johns Hopkins University Press, Baltimore, Maryland.

Hall, C.A.S., Cleveland, C.J. and Kaufmann, R. 1986. Energy and Resource Quality: The Ecology of the Economic Process. John Wiley & Sons, New York.

Hannon, B., Casler, S.D. and Blazeck, T.S. 1985. Energy Intensities for the U.S. Economy - 1977. University of Illinois, Urbana, Illinois.

Hilding, T. 1982. Gotländskt Vatten: Erfarenheter av Miljöstatistik på Regional och Lokal Nivå. Askö Laboratory, Stockholm University, Stockholm.

Jansson, A.M. 1988. The Ecological Economics of Sustainable Development: Environmental Conservation Reconsidered. In: The Stockholm Group for Studies on Natural Resources Management (eds.). Perspectives of Sustainable Development: Some Critical Issues Related to the Brundtland Report. Stockholm Studies In Natural Resources Management No 1. Department of Systems Ecology, Division of Natural Resources Management, Stockholm University, Stockholm. pp. 31-36.

Jansson, A.M. and Zucchetto, J. 1978. Energy, Economic and Ecological Relationships for Gotland, Sweden: A Regional Systems Study. *Ecological Bulletins* 28, Swedish Natural Science Research Council, Stockholm.

Jansson, A.M. and Titus, E. 1984. Agriculture and Water Quality in Lummelunda, Sweden. Askö Laboratory, Stockholm University, Stockholm. Mimeographed.

Jingsong, Y. and Honglu, Y. 1989. Integrated Fish Culture Management in China. In: Mitsch, W.J. and Jörgensen, S.E. 1989. Ecological Engineering: An Introduction to Ecotechnology. John Wiley and Sons, New York. pp. 375-408.

Lavine, M.J. 1984. Fossil Fuel and Sunlight: Relationship of Major Sources for Economic and Ecological Systems. In: Jansson, A.M. (ed.). Integration of Economy and Ecology: An Outlook for the Eighties. Proceedings from the Wallenberg Symposia. Askö Laboratory, Stockholm University, Stockholm. pp. 121-152.

Lavine, M.J. and Butler, T.J. 1981. Energy Analysis and Economic Analysis: A Comparison of Concepts. In: Mitsch, W.J., Bosserman, R.W. and Klopatek, J.J. (eds.). Energy and Ecological Modelling. Elsevier, New York. pp. 757-765.

Leontief, W.1966. Input-Output Economics. Oxford University Press, Oxford.

Lundqvist, G., Ernhold Hede, J. and Sundius, N. 1940. Beskrivning till Kartbladen Visby och Lummelunda. Sveriges Geologiska Undersökningar No. 183, Stockholm.

Lynne, G.D., Conroy, P.D. and Prochaska, F.J. 1981. Economic Valuation of Marsh Areas for Marine Production Processes. *Journal of Environmental Economics and Management* 8:175-186.

Mitsch, W.J. and Gosselink, J.G. 1986. Wetlands. Van Nostrand Reinhold, New York

Mitsch, W.J. and Jörgensen, S.E. 1989. Ecological Engineering: An Introduction to Ecotechnology. John Wiley and Sons, New York.

Mitsch, W.J., Reeder, B.C. and Klarer, D.M. 1989. The Role of Wetlands in Control of Nutrients with a Case Study of Western Lake Erie. In: Mitsch, W.J. and Jörgensen, S.E. 1989. Ecological Engineering: An Introduction to Ecotechnology. John Wiley and Sons, New York. pp. 129-158.

Nash, C. and Bowers, J. 1988. Alternative Approaches to the Valuation of Environmental Resources. In: Turner, R.K. (ed.). Sustainable Environmental Management: Principles and Practice. Belhaven Press, London. pp.118-142.

Nichols, D.S. 1983. Capacity of Natural Wetlands to Remove Nutrients from Wastewater. *Journal Water Pollution Control Federation* 55:495-505.

Nilsson, T. 1982. Markanvändning/Vattenkvalitet med Gotland som Testområde. Department of Hydrology, University of Uppsala, Uppsala, Sweden. Mimeographed.

Nordberg, M.-L. 1983. Markanvändning och Markanvändningsförändringar i Lummelundaåns Avrinningsområde, Gotland. Research Report 53. Department of Physical Geography, Stockholm.

Odum, E.P. 1979. Rebuttal of "Economic Value of Natural Coastal Wetlands: A Critique". *Coastal Zone Management Journal* 5:231-237.

Odum, E.P. 1981. A New Ecology for the Coast. Coastal Alert. Friends of the Earth for the Coast Alliance. pp. 146-181.

Odum, E.P. 1989. Ecology and Our Endangered Life-Support Systems. Sinuaer Associates, Sunderland, Massachusetts.

Odum, H.T. 1971. Environment, Power, and Society. John Wiley and Sons, New York.

Odum, H.T. 1984. Summary: Cypress Swamps and Their Regional Role. In: Ewel, K.C.

and Odum, H.T. 1984. (eds.) Cypress Swamps. University Presses of Florida, Gainesville. pp. 416-443.

Odum, H.T. and Odum, E.C. 1981. Energy Basis for Man and Nature. Second Edition. (Revised by H.T. Odum, August, 1987) McGraw-Hill, New York.

Östblom, G. 1986. Structural Change in the Swedish Economy: Empirical and Methodological Studies of Changes in Input-Output Structures. Dissertation for the doctor's degree in economics. Department of Economics, Stockholm University, Stockholm.

Pearce, D., Markandya, A. and Barbier, E.B. 1989. Blueprint for a Green Economy. Earthscan, London.

Petterson, B. 1946. Natur på Gotland. Bokförlaget Svensk Natur, Stockholm.

von Post, L. 1929. Om Gotlands myrar. Svenska Mosskulturföreningens tidsskrift 43:229-247.

Shijun, M. and Jingsong, Y. 1989. Ecological Engineering for Treatment and Utilization of Wastewater. In: Mitsch, W.J. and Jörgensen, S.E. 1989. Ecological Engineering: An Introduction to Ecotechnology. John Wiley and Sons, New York. pp. 185-217.

Spiller, G., Jansson, A.M. and Zucchetto, J. 1981. Modelling the Effects of Regional Energy Development on Groundwater Nitrate Pollution in Gotland, Sweden. In: Mitsch, J.W., Bosserman, R.W. and Klopatek, J.M. (eds.). Energy and Ecological Modelling. Elsevier Scientific Publishers, Amsterdam. pp. 495-506.

Stearns, F. 1978. Management Potentials: Summary and Recommendations. In: Good, R.E., Whigham, D.F. and Simpson, R.L. (eds.). Freshwater Wetlands: Ecological Processes and Management Potential. Academic Press, New York. pp. 357-363.

Thibodeau, F.R. and Ostro, B.D. 1981. An Economic Analysis of Wetland Protection. *Journal of Environmental Management* 12:19-30.

Turner, M., Odum, E.P., Costanza, R. 1988. Market and Nonmarket Values of the Georgia Landscape. *Environmental Management* 12:209-217.

Turner, R.K. 1988. Wetland Conservation: Economics and Ethics. In: Collard, D., Pearce, D. and Ulph, D. (eds.). Economics, Growth and Sustainable Environments. The Macmillan Press, London. pp. 121-160.

Westin, L., Nyman, L. and Gydemo, R. 1982. Gotländska Sjöar och Vattendrag: En Fiskeribiologisk Inventering. Information från Sötvattenslaboratoriet, No. 9. Drottningholm.

Whittaker, R.H. and Likens, G.E. 1975. (eds.). Primary Productivity of the Biosphere. Springer Verlag, New York.

Williams, E.M. and Eddy, F.B. 1989. Effect of Nitrate on the Embryonic Development of Atlantic Salmon (Salmo salar). *Canadian Journal of Fisheries and Aquatic Sciences* 46:1726-1729.

Zucchetto, J. and Jansson, A.M. 1985. Resources and Society: A Systems Ecology Study of the Island of Gotland, Sweden. Ecological Studies 56, Springer Verlag, Heidelberg.

Linking the Natural Environment and the Economy; Essays from the Eco-Eco Group,
Carl Folke and Tomas Kåberger (editors)
Second Edition.
1992. Kluwer Academic Publishers

CHAPTER 9

Costs for Nitrogen Source Reduction in a Eutrophicated Bay in Sweden

by

Ing-Marie Gren

Beijer International Institute of Ecological Economics
The Royal Swedish Academy of Sciences
Box 50005, S-104 05 Stockholm, Sweden

Eutrophication has wiped out the bottom fauna of Laholm Bay, on the Swedish west coast. Marine researchers have estimated a need for a 50 per cent reduction in the input of nitrogen in order to restore the bay. In this chapter cost-efficient reductions in the load of nitrogen are estimated. Measures to reduce the impact of nitrogen such as treatment of manure, fertilization, wetlands, sewage sludge and traffic emissions are evaluated. Localizations of nitrogen emissions are also taken into account. The study shows that a cost-efficient reduction in the supply of nitrogen by 50 per cent should be allocated among nitrogen sources such that agriculture would account for one half of the total reduction. Another important result is that redevelopment of wetlands is associated with low costs. The marginal cost for this measure amounts to only one third of the marginal cost of the second cheapest alternative.

1. INTRODUCTION

The Laholm Bay is located on the Swedish west coast about 150 km south of Gothenburg. Like many coastal waters around the world, the bay has experienced severe eutrophication problems for several years. Heavy phytoplankton blooms due to a high supply of nitrogen having caused oxygen deficits. Today, the bottom fauna in large areas of this water body is dead. In order to restore the biological conditions of the bay the total transport of nitrogen must be reduced by 50 per cent according to marine researchers. An interdisciplinary project of the drainage basin of Laholm, which included researchers from the sciences of marine biology, agriculture, economics and law, was therefore started to find measures leading to this reduction in the load of nitrogen. The purpose of this chapter is to present results from the economic part of the research where the minimum cost for reducing the supply of nitrogen by 50 per cent was estimated.

In this interdisciplinary project it was found that leaching from agriculture is high. The agricultural sector contributes to the load of nitrogen by the application of fertilizers and manure. Costs for measures involving the use of fertilizers and treatment of manure are therefore estimated, and this constitutes the main part of this chapter. Other sources of nitrogen also influence the water quality but probably to a lesser extent. In addition to agriculture, costs are calculated for nitrogen reductions by wetlands, from sewage treatment plants and traffic.

A large supply of nitrogen from arable land does not imply, however, that all available measures should be directed towards agriculture. A cost-efficient distribution of reductions between sources depends namely on two things *i)* the possible reductions in the load of nitrogen, and *ii)* the costs per unit of nitrogen reduction. When this information is available for each source of nitrogen emission we can determine the distribution of nitrogen reductions between sources which decrease the supply of nitrogen of Laholm Bay by 50 per cent at minimum cost.

Several studies have compared the costs of measures to reduce the application of fertilizers and leakage of manure, e.g. Andersson et al. (1987), Andréasson (1989) and Berger et al. (1989). Some studies have also included other pollution sources (e.g. Joelsson, 1986). To my knowledge, there is no study that has compared costs for all of the sources included in our study. It should, however, be noted that our aim of including all major sources of nitrogen is achieved at the expense of accuracy in the calculations of costs. For some measures we must rely on rough cost estimates. This is particularly true for the cost of reducing nitrogen emissions by improving the treatment of sewage discharge.

This chapter is organized as follows. We start with a brief description of the nitrogen sources and their emissions. Next, the model used to determine a cost efficient reduction in the nitrogen load is presented. In the third section we give results from the estimations of reduction costs and potential nitrogen reductions for all sources. Finally, the results are summarized and discussed. Throughout the paper, all numbers and calculations referred to are found in Fleischer (1989).

2. NITROGEN SOURCES AND THEIR EMISSIONS

The marine research of the south-eastern Kattegat identified two main sources of nitrogen; atmospheric deposition and the application of nitrogen on arable land. Nitrogen from the drainage basin is either transported to the bay by streams or discharged directly into the bay. A minor part of the air emissions is due to traffic and to the evaporation of ammonia.

From measurements in some streams, the maximum supply of nitrogen to Laholm Bay amounts to 6,620 tonnes of N (1977) and the minimum supply is 2,720 tonnes of N (1976). During the 1980's the transport of nitrogen has come close to the maximum supply. The load of nitrogen varies, not only between years, but also within a year. Most is transported by streams in the late summer and autumn. At this time of year the supply of nitrogen is especially damaging to water quality due to blooms of phytoplankton. Furthermore, a large part of the nitrogen is immediately available to the phytoplanktons; 52 per cent of the total supply of nitrogen is inorganic.

The total load of nitrogen to the bay from the drainage basin averages about 5,400 tonnes of N per year. Measurements were made for the period 1980-1987. Point sources account for 790 tonnes of N per year. The sewage treatment plants are the most important point sources. The non-point sources are dominated by the agricultural sector which emits almost 2,000 tonnes of N. All of the sources and their nitrogen emissions are listed in Table 1.

Table 1. Sources and annual supply of nitrogen to the Laholm Bay, 1980 - 1987.

Source	Tonnes of N	%
Agriculture	2,036	37.7
Forest areas, mires	2,207	40.9
Lakes	373	6.9
Industry	32	0.6
Sewage treatment plants	693	12.8
Private homes	59	1.1
TOTAL	**5,399**	**100.0**

Agriculture is undoubtedly the largest single source of nitrogen load to Laholm Bay. Emissions from forest areas include atmospheric deposition from the traffic. The third source of major importance to the water quality of the bay is sewage treatment plants. Costs for reductions in the supply of nitrogen from these three sources are therefore calculated in this chapter. The other sources are excluded since their nitrogen emissions are small and reduction costs are relatively difficult to estimate.

The location of the nitrogen emission source is of primary importance to the water quality of the bay. Emissions directly into the sea water are the most damaging. When the source is located further from the bay, its nitrogen emissions are less damaging to the quality of the water. This is due to the nitrogen retention which occurs during the transport to the bay.

The area of the drainage basin is approximately 10,100 km^2, most of which consists of land covered by forest. The acreage of arable land amounts to 12 per cent. However, the forests and crop fields are unevenly distributed in the drainage basin. The large areas covered by forest are located upstream while most of the arable land is found downstream. The drainage basin is therefore divided into three regions which are classified by their distance from the bay and their land use. The drainage basin of Laholm Bay and its subdivision into regions are shown in Figure 1.

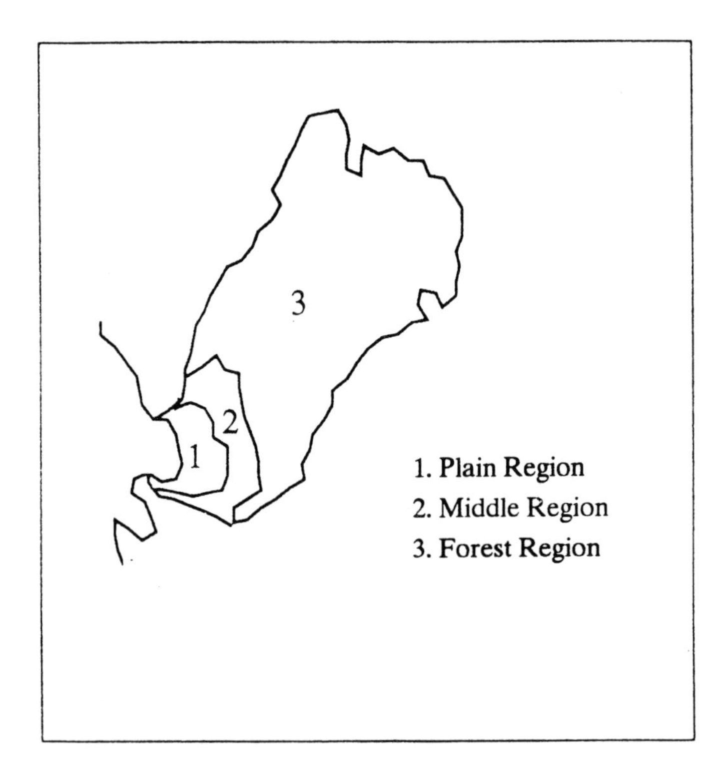

Figure 1. The regions of the Laholm Bay drainage basin.

In the region closest to the bay, region 1, grain production is intense. In this area the intensity of the nitrogen application is highest. In 1986, 180 kg N per hectare of artificial fertilizer and manure were applied on arable land. The middle region, region 2, is characterized by large holdings of livestock. Livestock units per hectare are twice as high as the average holding in Sweden. The application of nitrogen amounts to 74 kg N per hectare. The largest area of the drainage basin is region 3, where the acreage of forest land is high. In this area, the application of nitrogen is 46 kg N per hectare.

As mentioned above, some of the nitrogen applied to the land is retained during transport to the bay. The rate of retention differs between the three regions. Findings from the interdisciplinary project show that the retention factors are 0.15 in region 2 and 0.50 in region 3. In region 1, the coastal region, all nitrogen applied to the land leaches into the bay. These retention factors are used to calculate costs throughout this chapter.

THE MODEL

In the introduction it was stated that the purpose of this chapter is to find the distribution of nitrogen reduction between the sources which would give a 50 per cent reduction in the load of nitrogen to Laholm Bay at a minimum cost. This cost efficient distribution is determined by the possible reduction and marginal reduction cost of each source. Cost efficiency is achieved when certain conditions for optimum are met. These conditions are derived from the following problem of optimization, where total costs for reducing the supply of nitrogen to the bay are minimized. The costs are minimized subject to a 50 per cent reduction in the nitrogen supply and limited capacities of nitrogen reductions.

$$\text{(1)} \quad \min_{N^i} \sum_{i=1}^{n} C^i(N)$$

s.t

$$\sum_{i=1}^{n} \alpha^{ij} N^i = N^*$$

$$N^i < N^{io}$$

where $C^i(N)$: a cost function for each source i

N^i: nitrogen emission of each source i

N^*: desired total emission of nitrogen

α^{ij}: a distance coefficient for each region j, $0 < {}^{ij} < 1$

N^{io}: maximum nitrogen emission of each source i

At the optimum, the maximum nitrogen emissions, N^{i*}, are distributed between sources in such a way that the total costs are minimized. This is achieved when the following first-order condition for minimization holds

$$(2) \qquad \frac{\delta C^i/\delta N^i + \beta^i}{\alpha^{ij}} = \tau$$

where τ and β^i are the Lagrange multipliers of the restrictions. The economic interpretation of τ is that it measures the change in costs when the restriction on the desired load of nitrogen to the bay, N^*, is changed by one unit. The coefficient β^i measures the effect on total cost of expanding the maximum nitrogen reduction of a source. According to equation (2), cost efficiency is obtained when the marginal costs of nitrogen reduction are equal for all sources. Note that, due to the distance coefficient, it is more costly to obtain a reduction to the bay by, say, one kg of nitrogen from sources far away from the bay than from sources located in the coastal region. The distance coefficient, α^{ij}, is measured as one minus the retention rate, which gives $\alpha^{i1}= 1$, $\alpha^{i2}= 0.85$ and $\alpha^{i3}= 0.5$.

The implication of condition (2) can be clarified by an example. Assume that we have only two sources of nitrogen, A and B, and that they operate where marginal costs differ such that $MC^A= 20$ and $MC^B= 50$. We can then redistribute the reductions in the load of nitrogen between A and B such that the same total reduction is obtained at a lower cost. This is achieved by decreasing the reduction for B by one unit, which implies that costs are decreased by SEK 50. In order to maintain the total reduction in the load of nitrogen, the reduction is increased by one unit for A. Total costs are then increased by SEK 20, i.e. the net effect in total costs is a decrease by SEK 30.

The marginal costs for most measures regarding changes in the treatment of manure, energy forestry and reduced sewage discharge are calculated as the present values of investment costs. The same holds true for the costs of restoring wetlands where the investment cost is measured as the value of the alternative use of the land in question. Costs for cultivating new crops include the costs for labour and seed. Nitrogen from air emissions are assumed to be reduced by lowering the speed limits of the traffic. The cost for the nitrogen reduction is then calculated as the value of the extra time spent in driving a certain distance. In this study, the calculations of all these costs are much simplified. In general, the calculation of these costs is open to discussion, e.g. the evaluation of time savings in transport economics is subject to dispute and a large literature exists.

The cost of reducing the use of fertilizers is, however, a little bit more complicated to estimate. It is measured as the decrease in the net value of yield caused by a reduction in the application of nitrogen. It is then assumed that the revealed demand for nitrogen measures the marginal utility of yield at different levels of fertilizer application. The assumption is questioned since, in general, empirical studies of the agricultural sector do not include the effects of uncertainty. However, this assumption also holds true under some conditions of uncertainty, see Andréasson (1989). The derivation of changes in the value of yield by means of the demand for fertilizer is illustrated in Figure 2. For ease of illustration it is assumed that the demand function is linear.

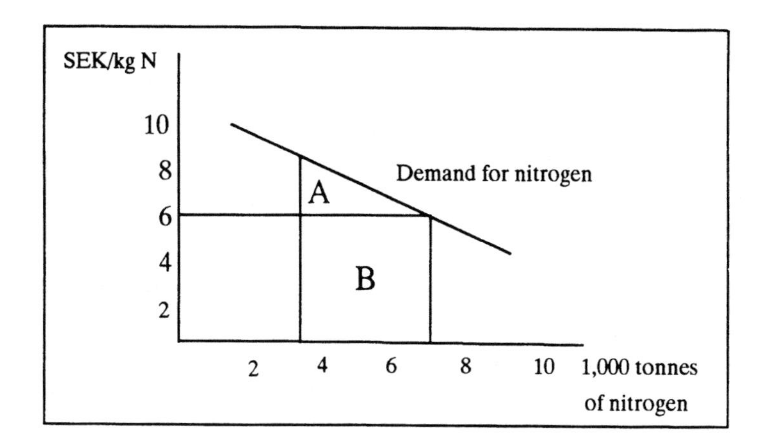

Figure 2. Value of yield loss and demand for nitrogen.

In 1986, the total use of fertilizer nitrogen in the drainage basin of Laholm was 7,200 tonnes of N and the price was SEK 6/kg N. This means that, at 7,200 tonnes, one additional kg of N increases the value of yield by SEK 6. The decrease in the net value of yield caused by a 50 per cent reduction in the use of nitrogen can then be estimated by means of the demand for nitrogen in the following way. Note that private and social costs for reducing the use of fertilizer differ due to subsidies on export of corn. This is accounted for in the empirical calculcations where all costs are measured at their social value.

When the use of nitrogen is reduced by 50 per cent the reduction in the value of yield corresponds to the area A+B in Figure 2. But, since less nitrogen is used costs are decreased corresponding to the area B in Figure 2. Thus, the change in the net value of yield is the area A, which we measure as the cost for reducing the use of fertilizer nitrogen by 50 per cent. In order to calculate this cost, the demand for nitrogen must be known. The demand for nitrogen is therefore estimated using econometric methods which are briefly described in the next section.

4. COSTS FOR DIFFERENT MEASURES

In the foregoing section it was shown that information on marginal costs for reducing the load of nitrogen is necessary for the determination of a cost efficient nitrogen reduction. In this section results from calculations of such marginal costs are presented. The measures included are; changes in the treatment of manure, cultivation of new crops, reduced application of fertilizers, restoration of wetlands, reduced sewage discharge and air emissions from traffic.

4.1 Changes in the Treatment of Manure and Cultivation of New Crops.

Measures regarding changes in the treatment of manure and cultivation of new crops are mainly aimed at reducing the leakage of nitrogen. Application of nitrogen in the autumn has the most severe effect on water quality since there are no crops available which can make use of the nitrogen. A reduction in the leakage can then be obtained either by changing the time when manure is spread from autumn to spring or by cultivating crops which can make use of the nitrogen applied in the autumn. Such crops included in this study are; cover crops, grass and energy forest. We also include reduction in the holdings of live-stock, which reduces the overall application of nitrogen. The marginal costs of these measures are listed in Table 2.

Table 2. Marginal costs for changes in the treatment of manure and cultivation of new crops.

Measure	SEK/kg Reduction of Nitrogen
Cover crops	12 - 55
Spreading in the spring instead of in the autumn	15 - 227
Energy forestry	26
Grass land	38
Reduced holdings of livestock	51 - 1954

The marginal cost of cover crops depends on in which region it is cultivated. It is cheapest to obtain a certain nitrogen reduction by cultivating cover crops in the region closest to the bay. This holds true also for change in spreading time and reduced holdings of livestock. However, the marginal costs of these measures also vary depending on the type of livestock. The costs for changing the use of manure were investigated for pigs, sows, cattle and poultry. It was found that it is least costly to change the spreading time for pig manure. It is also cheapest to reduce the holdings of pigs. For the measures without any variation in costs, i.e. for energy forestry and grassland, it is assumed that they are implemented only in the coastal region.

It should be noted that the calculation of costs for a change in the spreading time, reduced holdings of livestock and energy forestry implies discounting of future streams of costs and incomes. In subsequent sections we also have to discount future costs for reducing the nitrogen emissions from sewage treatment plants and for restoring wetlands. In order to convert future streams of payments into current values, the real discount rate is assumed to be 5 per cent. This rate is commonly used when discounting public projects in Sweden. However, one could argue in favour of both a lower and a higher level and a large literature exists on this topic.

4.2 Reduced Application of Fertilizers

The cost of reducing the application of nitrogen fertilizers is measured as the decrease in the net value of yield, as shown in the foregoing section. Information regarding the demand for nitrogen is necessary for determining this cost. Regression equations of the demand for nitrogen were therefore derived. As independent variables we included; price of fertilizers, wage rate, arable land, supply of manure and a time coefficient measuring monotonic technological growth.

Linear, logarithmic, exponential and hyperbolic demand functions were estimated. Due to measurement errors in the manure variable, we used the instrumental variable method to estimate all equations. The logarithmic regresstion equation turned out to give the best fit to our data. According to this equation, the nitrogen price elasticity is 0.35 in absolute value. This implies that the use of nitrogen is decreased by 35 per cent when the price of nitrogen is increased by 100 per cent. It should be noted that this result comes close to the values of nitrogen price elasticities found in several other fertilizer demand studies.

The marginal costs of reducing the application of fertilizers are sensitive not only to the price elasticity but also to the rate of nitrogen leakage. Unfortunately, no experiments have been carried out for the drainage basin of Laholm. We must therefore use results from experiments carried out in other parts of Sweden. According to these results, the nitrogen leakage from fertilizers varies between 5 and 60 kg N/ha. The nitrogen leakage depends on, among other things, the application rate of manure. If the application of manure remains unchanged, the nitrogen leakage of fertilizers is probably high. When the leakage and application of manure is reduced the nitrogen surplus in soil decreases. This implies that crops can fully utilize the fertilizer applied in spring and summer time. We would then expect the nitrogen leakage of fertilizers to be rather low and we assume that the leakage is 15 kg N/ha.

Given the above assumptions, marginal costs for reducing the use of fertilizers range between SEK 14/kg N and SEK 348/kg N. If no fertilizers were applied on the fields, the load of nitrogen to the bay of Laholm would be reduced by about 800 tonnes.

More than one half of this reduction is obtained by stopping the fertilizer application in the coastal region. This is mainly due to our assumptions that all nitrogen applied in the coastal region leaks into the bay while only part of the nitrogen applied in more remote regions reaches the bay.

4.3 Traffic

Nitrogen oxide emissions from traffic are assumed to be reduced by a decrease in the speed limits from 110 km/h or 90 km/h to 70 km/h. About 40 per cent of the air emissions are due to the traffic on the main roads where the current speed limit is 110 km/h. The cost of reducing the air emissions is calculated as the loss of time due to the lower speed limit. It should be noted that we do not include other benefits from lowering the speed limits, such as a lower accident rate and reduced emissions of carbon monoxide.

The costs for reducing the air emissions by lowering speed limits thus depend on the valuation of time, which is wide open for discussion. In our calculations we used values estimated by a Swedish department where decisions are made regarding investments in the road network. Different kinds of travel for leisure and work are accounted for and the average value is then SEK 40/hour.

The results from our calculations show that the marginal costs for reducing the load of nitrogen to Laholm Bay by lowering the speed limits vary between SEK 194/kg N and 1,616/kg N. The maximum reduction in the load of nitrogen is 325 tonnes of nitrogen. Note that the marginal costs are high as compared to the marginal costs of the measures described above.

4.4 Sewage Discharge

In the drainage basin of Laholm there are altogether 55 sewage treatment plants. In order to obtain accurate estimates of the marginal costs for reducing the nitrogen discharge, all these plants should be investigated. This has not been possible within this project. Instead, we rely on some rough measures valid for an average Swedish plant.

On average, the nitrogen discharge is reduced by 30 per cent in the drainage basin of Laholm. It is assumed that nitrogen reduction is increased by installing extra treatment beds in all existing treatment plants. The marginal cost for increasing the rate of nitrogen reduction to 50 per cent then varies between SEK 50/kg N and SEK 100/kg N depending on in which region the plant is located. The marginal cost for increasing the nitrogen reduction from 50 to 75 per cent ranges between SEK 100/kg N and SEK 200/kg N.

4.5 Restoration of Wetlands

One way to decrease the load of nitrogen to Laholm Bay is to increase the retention of nitrogen during the transport to the coastal water. This is obtained by restoring wetlands, especially in the coastal region. Unfortunately, very little is known about the effects of such restoration since very few large scale experiments have been carried out. There are, however, some results available from Danish and U.S. experiments which show that the leakage of nitrogen could be reduced by approximately 800 kg N/ha.

The cost for restoring wetlands is calculated as the value of the alternative use of the land in question. The value of land varies depending on the quality of soil etc. No studies have been made on which areas are most appropriate for the creation of wetlands in the drainage basin. We therefore assume that the farmers are compensated by an amount corresponding to the value of the yield from high-productive crop land, which amounts to about SEK 3,500/ha per year. The constant marginal cost for reducing the load of nitrogen to Laholm Bay by restoring wetlands is then SEK 4/kg N.

Since the cost per unit of nitrogen is low as compared to other measures, the least costly way of obtaining the necessary reduction of nitrogen to the bay would be to convert the required area of arable land into wetlands. There is, however, no legal support for this measure. If the farmers do not sell the land required, the use of wetland is restricted by the available area of arable land. We thus assume that wetlands can be created on 500 ha of the arable land.

5. NITROGEN REDUCTIONS AT MINIMUM COSTS

We know from the analytical section that cost efficiency is determined by marginal costs and potential reductions in the load of nitrogen for each source. Results from calculations of marginal costs and potential nitrogen reductions were presented in the two preceeding sections. These results are summarized in Table 3.

Restoration of wetlands is undoubtedly the least costly measure. The marginal cost amounts to SEK 4/kg N while the marginal cost of the second cheapest measure, cultivation of cover crops in the coastal region, is SEK 12/kg N. In general, we find that it is possible to use relatively cheap measures within the agricultural sector such as reducing the application of fertilizers, change in spreading time for manure and energy forestry. But the most expensive measure, reduced holdings of livestock, also belongs to the agricultural sector. Thus, we conclude that not all measures involving agriculture should be enforced when reducing the supply of nitrogen by 50 per cent, i.e. 2,700 tonnes of nitrogen, at minimum costs.

In order to find the minimum cost of reducing the supply of nitrogen by 2,700 tonnes of nitrogen the values in Table 3 were used for calculating the costs for various nitrogen reductions. The result gives the minimum costs for different nitrogen reductions, and this is illustrated in Figure 3.

Table 3. Marginal costs and potential nitrogen reductions.

Measure	SEK/kg N	Tonnes of N
Wetland	4	430
Agriculture		
Cover crops	12 - 55	75
Fertilizers	14 - 348	798
Spreading of manure in autumn instead of in the spring	15 - 227	190
Energy forestry	26	300
Grassland	38	109
Reduced holdings	51 - 1954	150
Sewage discharge	50 - 200	600
Traffic	194 -1616	325
Total	**4 -1954**	**2977**

The curve in Figure 3 shows the minimum cost for reducing the load of nitrogen to Laholm bay by different quantities. Moving along the curve we can see that the total minimum cost for reducing the nitrogen supply by 50 per cent, i.e. 2,700 tonnes, amounts to SEK 175 millions per year. If all possible reductions were enforced the total minimum cost would be SEK 320 millions per year. A reduction corresponding to 55 per cent of the total supply would then be achieved.

It was mentioned above that a cost efficient reduction in the supply of nitrogen by 50 per cent requires enforcement of all the measures listed in Table 3, although to different degrees. Since restoration of wetlands is much cheaper than all other measures, this measure should be used as much as possible. Speed limits for traffic should however be implemented to a limited extent; see Table 4.

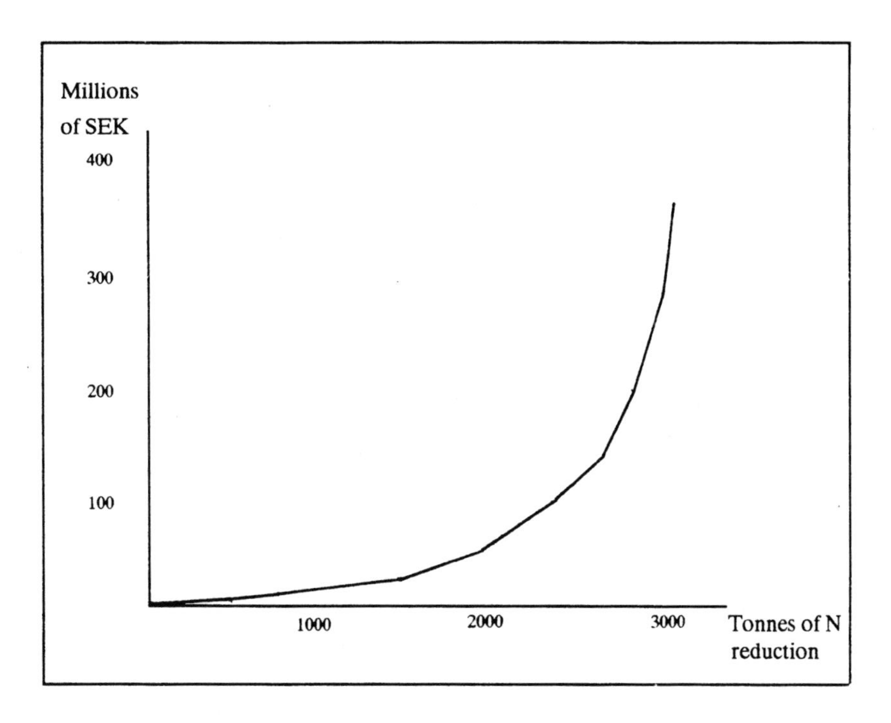

Figure 3. Minimum costs for reducing the nitrogen supply to Laholm bay.

Agriculture accounts for the largest part of the total nitrogen reduction and costs. Notice the large diffences between the shares of reductions and costs for restoration of wetlands. The share of the total reduction is 16 per cent while the share of costs amounts to only 1 per cent. The reverse is true for traffic emission. Reducing the nitrogen supply by this measure is relatively expensive.

Remember from the description of costs for different measures that no benefits associated with any measure are included. This might be of considerable importance for reduction of speed limits since then not only emissions of nitrogen oxides are reduced but also the emissions of carbon-monoxides. Furthermore, the accident rate and wear of roadways are reduced. If these factors were included in our calculations of costs, traffic would probably account for a much larger part than is shown in Table 4. The cost for restoring wetlands would also be reduced if the benefits were included. Wetlands perform several functions, such as cleaning and storing of water, which are of great value to humans. The inclusion of these benefits might turn the cost of wetlands into a net benefit.

Table 4. Distribution of nitrogen reductions and costs for a 50 per cent reduction in the nitrogen supply to Laholm Bay.

	Agri-culture	**Wetlands**	**Sewage discharge**	**Traffic**	**Total**
Tons of N	1615	430	600	55	2700
Per cent	60	16	22	2	100
Millions of SEK	91.9	1.7	65.9	16.2	175.7
Per cent	52	1	38	9	100

It should also be noted that the results in Table 4 are sensitive to the assumptions made regarding the nitrogen leakage of fertilizers and appropriate land available for restoring wetlands. Remember that we assumed the nitrogen leakage of fertilizers to amount to 15 kg N/ha. If the real leakage is higher, say 20 kg N/ha, total cost for reducing the leakage to the bay by 2,700 tonnes is reduced from 176 millions of SEK to approximately 120 millions of SEK. Sensitivity analysis also shows that total costs are very sensitive to the area of land available for restoring wetlands. The costs would be reduced to about 100 millions of SEK if the available land was twice as large, i.e. 10 km^2 instead of 5 km^2.

Another factor which is not taken into account is the time period required for a measure before it reaches full effect. Once extra nitrogen reduction beds are installed in existing sewage works, a reduction in the discharge of sewage into the water has an immediate effect on the water quality. On the other hand, a reduction in the application of nitrogen improves the water quality after several years due to the storage of nitrogen in soil. An adjustment of the cost with respect to the time perspective would propably increase the costs for some measures within the agricultural sector, e.g., reduction of fertilization and decrease of the holdings of livestock.

6. SUMMARY OF THE RESULTS

The purpose of this chapter has been to present calculations of costs for measures aimed at reducing the supply of nitrogen to the Laholm bay. The bottom fauna in this water body is now dead due to oxygen deficiency caused by a high supply of nitrogen.

According to marine research, the load of nitrogen must be reduced by 50 per cent in order to restore the biological conditions. In an interdisciplinary project, it was found that agriculture is the largest single source of nitrogen in the drainage basin. Other nitrogen sources included in this study were: wetlands, sewage discharge and traffic emissions. The fact that the impact on the Laholm Bay of a certain amount of nitrogen depends on the localization of the deposition was also taken into account.

In order to find a cost efficient reduction in the supply of nitrogen we must, for each nitrogen source, know *i)* maximum reduction in the emission of nitrogen and *ii)* marginal cost for nitrogen reduction. It was found that a maximum reduction of 3,000 tonnes of nitrogen can be achieved. This is slightly more than the reduction required to restore the bottom fauna of the bay which amounts to 2,700 tonnes. Measures involving the use of fertilizer and treatment of manure account for about one half of the total potential reduction in the supply of nitrogen to the bay.

The calculations showed that the minimum total cost for reducing the load of nitrogen to the Laholm bay by 50 per cent, i.e. 2,700 tonnes, amounts to approximately SEK 175 millions per year. The agricultural sector then accounts for the largest part, 60 per cent, due to its large share of potential total reduction and relatively low reduction costs. It turned out, however, that the undoubtedly cheapest measure is restoration of wetlands. The cost for this measure is SEK 4/kg reduction of nitrogen, while the cost of the second cheapest source, cultivation of cover crops, amounts to SEK 12/kg.

It shold be noted that the calculations of costs are based on several assumptions regarding: nitrogen leakage of fertilizers, appropriate land available for restoring wetlands and traffic emissions. The figures must therefore be taken with caution. But, nevertheless, robust results are that agriculture must account for the largest part of the total nitrogen reduction and that restoration of wetlands turned out to be a cheap measure which deserves further research.

REFERENCES

Andersson, A-K. and Svanäng, K. 1987. Three Economic Studies of Water Treatment Within the Agricultural Sector. Department of Economics, Swedish University of Agricultural Sciences, Uppsala.

Andréasson, I-M. 1989. Costs of Controls on Farmers' Use of Nitrogen - A Study Applied to Gotland. Dissertation for the doctor's degree in economics. The Economic Research Institute, Stockholm School of Economics, Stockholm.

Berger, M. S. and Johnsen, F.H. 1989. Cost-effectiveness Analysis of Measures Against Agricultural Pollution of Lake Mjösa. In: Dubgaard, A. and Nielsen, A.H. (ed.).

Economic Aspects of Environmental Regulations in Agriculture. Wissenschaftsverlag Vauk Kiel KG.

Fleischer, S. (ed.). 1989. Water Quality and Land Use. - A Study of the Drainage Basin of Laholm. Report No. 1. County Board of Halland, S - 301 86 Halmstad.

Joelsson A. (ed.). 1986. Measures For Reducing the Supply of Nitrogen and Phosphorus to the Bay of Laholm. County Board of Halland, S - 301 86 Halmstad.

Linking the Natural Environment and the Economy; Essays from the Eco-Eco Group,
Carl Folke and Tomas Kåberger (editors)
Second Edition.
1992. Kluwer Academic Publishers

CHAPTER 10

Marine Ecosystem Support to Fisheries and Fish Trade

by

Monica Hammer

Department of Systems Ecology
Stockholm University
S-106 91 Stockholm, Sweden

The relationships between fish production in marine ecosystems and fish processed and traded in the economy is analyzed for Sweden's foreign trade in herring, cod, salmon and fish meal for 1986. Estimates are made of direct and indirect energy requirements in the marine ecosystem and the economy to produce the traded fish products. A comparison is made of trade balances in economic terms, expressed as export and import prices and in energy terms, expressed in solar energy terms. In monetary terms imports are 2.5 times larger than exports, but in energy terms about 7.3 times larger than exports meaning that Sweden is receiving products and work performed in the marine ecosystems at a lower price than it is selling its own products. This indicates that Sweden is dependent on much larger ecosystem support areas performing necessary work than is reflected in standard economic evaluation. The importance of considering such ecosystem support in economic decision-making and management of indigenous and foreign living marine resources is emphasized.

1. INTRODUCTION

The intensive development of the commercial sea fisheries in recent decades and the resultant overfishing problems in several regions has increased awareness of the carrying capacity limits of the marine ecosystems (Harden-Jones, 1974; Cushing; 1982, Gulland, 1984; Hall et al., 1986). The development of world markets for fish products and intensified trade have also globalized fish resources.

Being a part of nature, man is dependent on the functioning of a number of ecosystem life-support processes. The economic system can be regarded as a subsystem of the overall ecosystem (Daly, 1987) where natural resources are brought into the economic system as they are exploited. However, the work done by marine ecosystems required to produce the exploited resource such as fish are seldom explicitly expressed in economic analyses. In managing fisheries, the primary emphasis has usually been the fish stocks of commercial species isolated from their supporting ecosystem. Since the outcome for the fisherman is directly connected to the natural production of the marine ecosystem, the strong interdependences between the functioning of the ecosystem and the opportunities for fish exploitation has been clearly demonstrated by the many examples of overfishing and degradation of fish resources (Harden-Jones, 1974; May, 1984).

Since the 1960's, Sweden has been a net importer of fish products, especially fish meal. A general opinion among different representatives in the Swedish fisheries and aquaculture industry has been that Sweden should decrease its dependence on imports and instead develop the domestic fishery industry (Anon., 1988; 1989). In this chapter I examine Sweden's trade balance with an ecological support perspective by comparing export and import prices in Swedish fish trade in 1986 with the direct and indirect energy required to produce the traded fish products. Energy cost is used here as an estimate of the biophysical[1] requirements in both the ecosystem and the economy.

2. FISHERIES AND FISH TRADE IN SWEDEN

Sweden's commercial fisheries are based on a few fish species, with herring (Clupea harengus) and cod (Gadus morhua) accounting for 80 per cent of the total catch. Approximately 4,500 Swedish fishermen harvest 200,000 tonnes of fish per year (Anon., 1987c). The waters off the 2,500 km long Swedish coastline (Anon., 1984) now represent the resource base for the sea fisheries with catch quotas (TAC) regulating harvest (Figure 1). The North Sea used to be the dominant fishing area for Sweden, but the population crashes of Atlanto-Scandic herring stocks and the new jurisdictional regulations of the seas (Law of the Sea, 1982), have shifted fishing pressure to the Baltic Sea, where half of Sweden's total catch is now taken.

A highly mechanized offshore fishery using trawlers accounts for around 80 per pent of the total catch. Alongside the offshore fishery, there exists a small scale coastal fishery catching a variety of species. Especially on the Baltic coast, fishing is a subsidiary industry to agriculture, forestry and other means of livelihood.

Herring catches normally exceed domestic demand while the catches of cod, flatfish, shellfish and scrap fish used in fish meal production have been insufficient for Swedish demand. In 1986, Sweden had a total import surplus in fish products of 2 times

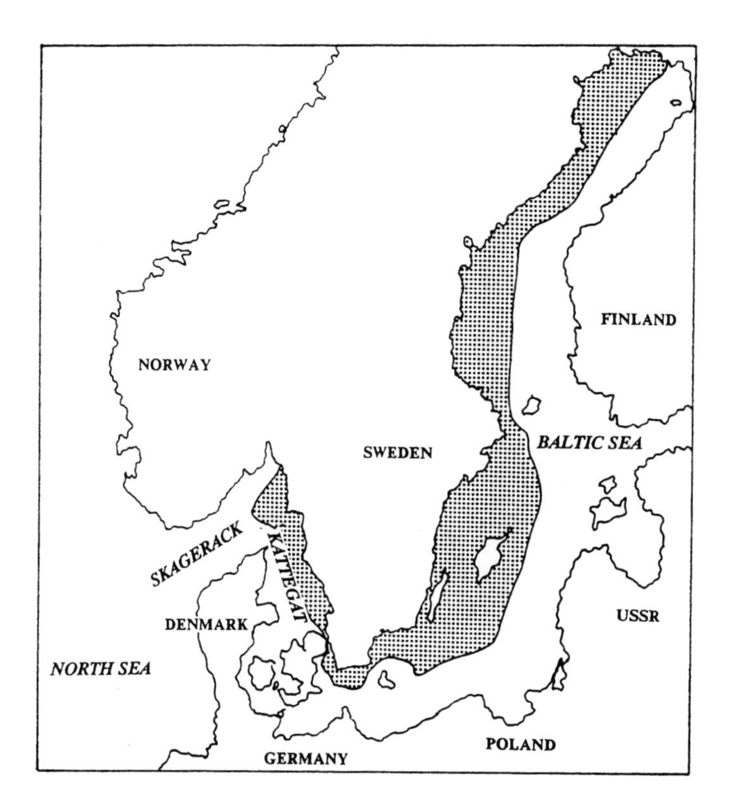

Figure 1. Sweden's fishing areas.

in product volume and 3.4 times in money value (Anon., 1987a). My analysis of the 1986 fish trade concentrates on herring, cod and salmon products and fish meal which together represented 57 per cent of the imports and 55 per cent of exports in tonnes, and 32 per cent and 42 per cent respectively in monetary terms.

Most of the imported fish products come from Norway and Denmark, countries which fish mainly in the North Sea, the Baltic Sea and Barents Sea. Salmon is to a large extent imported from Canada and USA (70 per cent), and some fish meal is imported from Iceland.

Major species used in fish meal production are herring, sprat, sand eel, blue whitling, capelin, Norway pout, and mackerel (Anon., 1986a; Anon., 1987a).

Table 1 summarizes Sweden's exports and imports of herring, cod, salmon and fish meal in 1986. Processed fish aimed for direct human consumption represents around 11 per cent in weight both of exports and imports. In imports, fish meal is the dominant product, accounting for 83 per cent by weight and 44 per cent by monetary

Table 1. Swedish imports and exports of herring, cod and salmon fish products and fish meal in 1986 expressed in tonnes and Swedish Crowns (SEK) (Anon., 1987b).

	Imports				Exports			
	ton 10^3	per cent	SEK 10^6	per cent	ton 10^3	per cent	SEK 10^6	per cent
unprocessed[a]								
herring	0.4	0.3	1.2	0.2	24.4	51.8	64.5	21.6
cod	0.3	0.2	2.7	0.4	14.1	29.9	107.4	36.0
salmon	7.4	5.6	204.1	27.0	0.4	0.8	14.3	4.8
total	**8.1**		**208.0**		**38.9**		**186.2**	
processed								
herring	8.2	6.3	66.8	8.9	3.4	7.2	53.2	17.8
cod	6.0	4.6	133.8	17.7	1.8	3.8	35.1	11.8
salmon	0.2	0.2	14.4	1.8	0.1	0.3	15.9	5.3
total	**14.4**		**215.0**		**5.8**		**104.2**	
fishmeal	108.6	82.8	331.7	44.0	2.9	6.2	8.0	2.7
Total	**131.1**	**100.0**	**754.7**	**100.0**	**47.1**	**100.0**	**298.4**	**100.0**

[a] As unprocessed fish is considered whole fresh and frozen fish for human consumption.

value. Sweden presently imports around 90 per cent of its total demand of fish meal (Anon., 1987d). In exports, on the other hand, the unprocessed fish for direct human consumption is the dominant product with 83 per cent by weight and 62 per cent in monetary value.

3. ENERGY ANALYSIS

Energy analysis focuses on energy flows with one of its goals being to estimate the direct and indirect energy needed to generate a commodity or maintain a process (Costanza, 1980; Lavine & Butler, 1982). Using energy as a common denominator allows comparisons between economic and ecological systems since energy flows through both systems (Odum, 1983).

A variety of different techniques are used in energy analysis. Some focus on the auxiliary energy use in the economy (electricity, fossil fuels etc.) to produce goods and services, such as the input-output method developed by Herendeen and Bullard (1974). Others consider food-chain based energies, e.g. Costanza (1980) who expanded the boundaries of the economic output-input matrix to include the final demand sector (government and households) and both solar and fossil fuel energy. A third approach uses solar energy equivalents as a common denominator. Here the concept of energy quality[2] is crucial for the analysis using transformation ratios derived from the relationship between the solar insolation and the total workings of the biosphere (Odum, 1987)

In this study of the Swedish fish trade, I estimate the flows of solar energy required in the marine ecosystem to produce the fish biomass needed to support the fish trade. Solar energy, fixed in primary production and concentrated through the food-chain drives ecosystem life processes. The capacity of marine ecosystems to produce fish is ultimately determined by the fixation of solar energy by algae and plants (Jansson, 1989). Nixon et al. (1986) demonstrated the relationship between fish yield and net primary production. Estimates based on landings data and reported net primary production from various marine systems including the Baltic Sea and the North Sea showed that overall efficiency lies between approximately 0.1 and 1 per cent.

The direct and indirect auxiliary energy inputs in fisheries and the fish processing industry are estimated in fossil fuel equivalents. A transformation factor of 10 was used to compare fossil and solar energy values (Odum, 1983; Boynton et al., 1984).

The term embodied energy in this study refers to the direct and indirect energy required to sustain production of a certain commodity whether in the economy or in the ecosystem. A conceptual model identifying the major flows of energy and materials in the marine ecosystems as well as in fisheries and fish processing industries is presented in Figure 2.

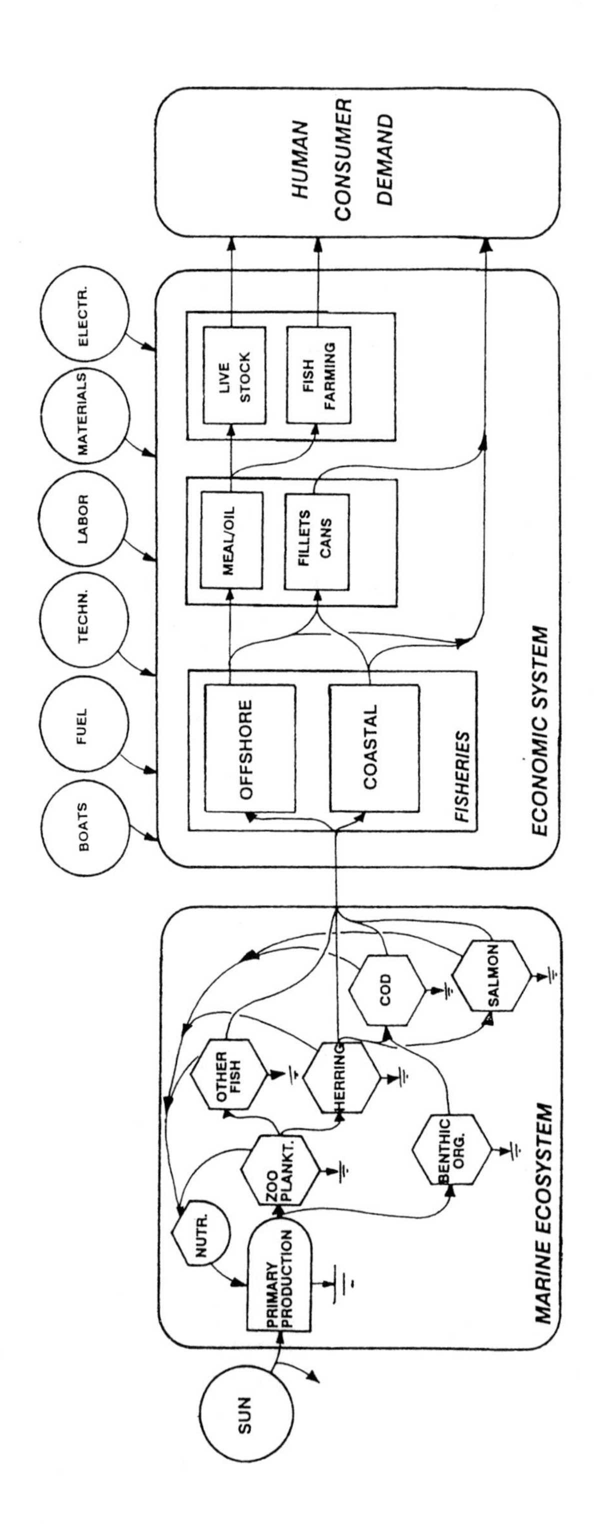

Figure 2. Major flows of energy and materials in ecological and economic systems involved in production of various fish products.

4. ENERGY INPUTS FROM THE MARINE ECOSYSTEM

The quantities of traded fish products of herring, cod, salmon and fish meal recorded in official statistics (Anon., 1987a; Anon., 1987b) are used as initial values for calculating the direct and indirect energy requirements in the marine ecosystem to produce the harvested fish biomass. Corrections for weight losses in processing are made. To produce 1 kg of fish meal, 5-6 kg of raw fish is needed, while in fillet processing 2-3 kg raw fish per kg fillet is required (Leach, 1976; Anon., 1985; Anon., 1986a; Limburg, 1986).

Calculations of the amount of solar energy fixation needed to support fish production is derived from existing data on fish production and primary production. This method assumes that the whole food web is necessary for the production of the fish. This may overestimate energy costs, but it does account for feedback flows in the food web. However, other environmental functions necessary for ecosystem production such as winds, waves and currents are not included (Jansson & Jansson, 1988).[3]

Estimates from the Baltic Sea, by Elmgren (1984), are used in this study as an approximation of primary production requirements for fish production. His estimates give an efficiency between fish yield and primary production of approximately 0.1 per cent which is in the lower part of the efficiency range of 0.1 - 1 per cent estimated by Nixon et al. (1986).

5. ENERGY INPUTS IN FISHERIES AND PROCESSING

Besides the direct inputs of fuels (gasoline and diesel oil), energy is required for building boats, gear and other technical devices used, as well as for the production of ice, salt and other necessary input materials. In addition, the energy cost of labor has to be considered.

Leach (1976) made the following estimates of the energy inputs in UK fisheries derived from UK input-output tables: Shipbuilding, equipment and ice amounted to less than 4 per cent of the total inputs of energy (labor excluded). In Japanese fisheries, Watanabe and Okubo (1989) found that direct fuel inputs varied from 77 per cent in salmon gill net fishery to 92 per cent for large North Pacific trawlers. In North East fisheries in the USA, direct fuel costs of operation dominate total energy costs for all vessel sizes from around 7 GRT to 330 GRT. However construction costs became more important with increasing vessel size (Pimentel & Pimentel, 1979). In Swedish and Danish fisheries, less than 2 per cent of the fishing vessels over 10 GRT are larger than 250 GRT, while Norway´s fishing fleet to a higher degree consists of large vessels including purse seiners and mother ship trawlers (Anon., 1986c; Anon., 1987e; Anon., 1987f). Based on this information, I assume that direct fuel alone is an approximate but accurate indication of the total energy input in fishery operations.

Fuel use depends largerly on fishing techniques, abundance of fish and distance to fishing grounds. The fishery is divided into offshore fishery using trawling or purse seines and coastal fishery.

The energy cost of labor in fisheries is estimated by using the relation between total energy use in Sweden and Gross National Product (GNP) (Jansson & Zuchetto, 1978). This method assumes a strong relationship between total auxiliary energy use and GNP (Cleveland et al., 1984; Hall et al., 1986). This approach represents not only the food intake of the worker but the energy support for a particular standard of living in the economy.[4]

Estimates of energy inputs in fish processing were derived from the existing literature. Processing fish meal in a 50-60 tonnes/day unit, which corresponds to the Swedish size of meal factories. Leach (1976) uses 11.8 GJ/tonne for inputs of steam and electricity. Lorentzen (1978) estimated energy inputs in Norwegian fillet processing (electricity, materials, depreciation) to be 13 GJ/tonne.

Labor inputs in fish fillet processing was estimated using an efficiency of approximately 42 tonnes of raw fish per laborer annually (Limburg, 1986). In Swedish fish meal processing around 550 - 700 tonnes of meal per laborer annually is produced (Christensen, S.I., Ängholmen's fish meal factory, personal communication).

A more detailed presentation of collected data and calculations is found in Hammer (1990).

6. RESULTS

Sweden's imports and exports for herring, cod, salmon and fish meal are quantified in four ways; by monetary value of traded products; by weight of traded products; by required solar and auxiliary energy; and by sea surface area needed to support the underlying ecosystem.

The monetary value of traded products, as recorded in official trade statistics, shows that imports are 2.5 times larger than exports. The weight of the traded fish products shows a slightly larger import surplus of 2.8.

When the work done by ecosystems is accounted for by estimating the required solar and auxiliary energy, the import surplus increases substantially to 7.1. The corresponding sea surface area needed to produce the imports is 8.3 times larger than that needed for exports.

The results are presented in Figure 3 and in Tables 2 and 3. Figure 3 shows the energy input, expressed in solar energy equivalents, and the money paid for exports and imports. As can be seen from the figure, a large amount of work is performed in the ecosystem, in terms of solar energy, before the fish is harvested. Expressed in units of similar quality, the solar energy inputs from the marine ecosystem are 6 times higher

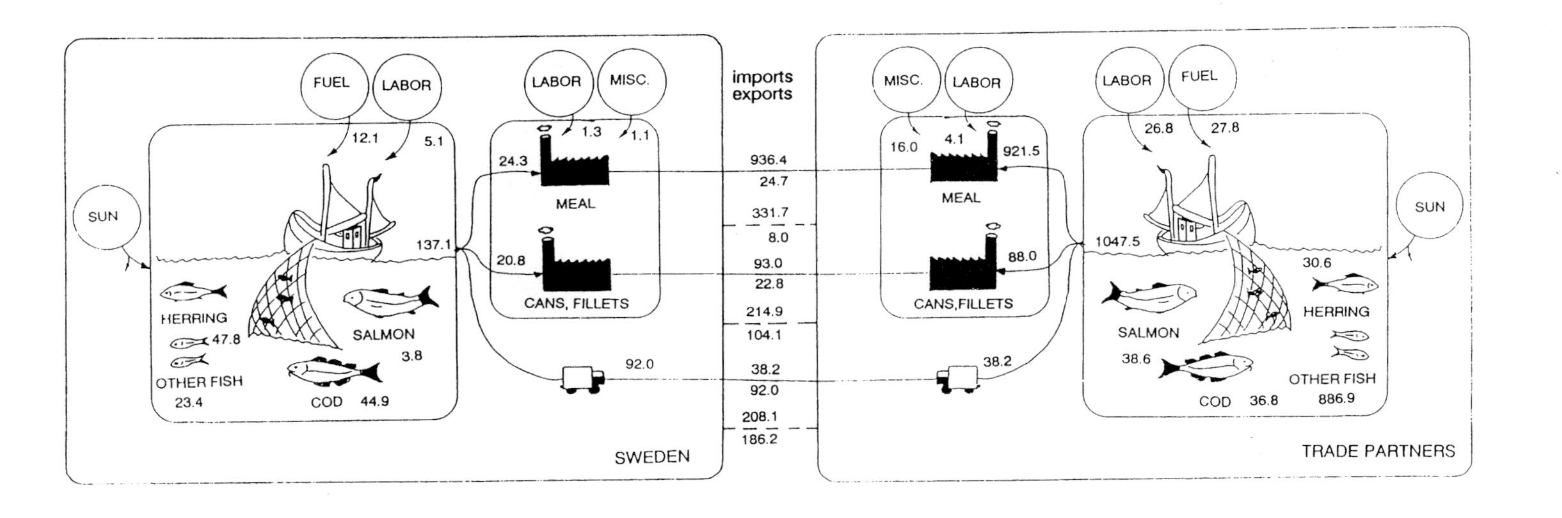

Figure 3. Direct and indirect energy costs in Swedish foreign fish trade in 1986, expressed in PJ solar energy equivalents (——). Corresponding export and import prices expressed in million Swedish Crowns (-- -- --).

Table 2. *Swedish exports and imports in herring, cod, salmon, and scrapfish (used for fish meal) in 1986 expressed in tonnes harvested biomass (fresh weight) and fixed solar energy (primary production). Inputs from the economy are expressed as direct and indirect auxiliary energy cost (in solar energy equivalents).*

	Natural Ecosystem				Fisheries		Processing	
	Traded goods (1000 ton)	**Fish harvest[a] (1000 ton)**	**Primary prod.[b] (PJ)**	**Prod. area[c] (km^2)**	**Fuel[d] (PJ)**	**Labor[e] (PJ)**	**Misc.[f] (PJ)**	**Labor[g] (PJ)**
Imports								
herring	8.6	21.0	30.6	5,271	4.5	1.5	1.1	1.8
cod	6.3	17.7	36.8	6,349	3.6	1.3	0.8	1.5
salmon	7.6	8.6	38.6	6,661	1.8	7.2	<0.1	<0.1
scrapfish	108.6	597.4	886.9	152,900	17.9	16.7	14.1	0.7
Total	**131.1**	**644.7**	**992.9**	**171,200**	**27.8**	**26.8**	**16.0**	**4.0**
Exports								
herring	27.8	32.9	47.8	8,245	7.0	2.4	0.4	0.7
cod	15.9	21.5	44.9	7,737	4.4	1.6	0.2	0.5
salmon	0.5	0.8	3.8	648	0.2	0.7	<0.1	<0.1
scrapfish	2.9	15.8	23.4	4,039	0.5	0.4	0.4	<0.1
Total	**47.1**	**71.0**	**119.9**	**20,690**	**12.1**	**5.1**	**1.1**	**1.2**

Notes to Table 2.

a) Weight losses before landing was accounted for by using conversion factors from Statistics Sweden (Anon,1987a). Raw fish required in fillet processing 2.5 times output (Anon., 1985; Alexandersson, Findus Bjuv AB, pers. comm.; Limburg, 1986) and in fish meal production 5.5 times (Anon., 1986a; Christensen, Ängholmens Fish meal factory, pers. comm.; Leach, 1976).

b) Primary production (PP) requirements for harvested fish, estimated from relation between Baltic Sea average PP (139 g $Cm^{-2}yr^{-1}$) and fish production (0.5 g $Cm^{-2}yr^{-1}$ for herring and scrap fish species and 0.2 g $Cm^{-2}yr^{-1}$ for cod and salmon). Data from Elmgren (1984).

c) Total primary production divided by the average primary production of 139 g C m^{-2} yr^{-1} = 5.8 MJ m^{-2}.

d) Direct fuel inputs 3.0 GJ/tonne for large purse seine (scrapfish), 24 GJ/tonne for offshore trawling, and 10 GJ/tonne for coastal fishery (Leach, 1976; Anon, 1988).

e) Total energy use in $Sweden_{1986}$ /GNP_{1986} =4.231 MJ/SEK; average salary, 8,500 SEK/month gives 432 GJ/fisherman /yr.

f) 13 GJ/tonne, refers to direct electricity, materials, depreciation estimated for Norwegian fillet processing (Lorentzen, 1978; see also Leach, 1976; Pitcher, 1977).

g) see e) 42 tonnes of raw fish processed/laborer/year in fillet processing and 650 tonnes of meal/laborer/year; 650 tonnes of meal/pers/yr (Limburg, 1986; Christensen, Ängholmens Fish meal factory, pers. comm.).

For further details, see Hammer (1990).

in exports and about 13 times higher in imports than the inputs of auxiliary fossil energy from the economy, even though exports are dominated by fresh and frozen (unprocessed) fish, as opposed to imports which are dominated by processed fish products.

Around 95 per cent (in energy terms) of the fish that is imported to Sweden is first processed to fish meal or to products for direct human consumption (fillets, cans). Fish meal is the main product representing around 88 per cent of the harvest. In exports, on the other hand, almost 80 per cent of the harvest is directly exported as whole fresh or frozen fish.

Since the raw fish requirement in fish processing for direct human consumption are around 2.5 times per unit final product produced (fillets) while fish meal, which is dried, needs around 5.5 times the output, the total fish biomass that it is necessary to harvest in order to compensate for the imports is considerably larger than that required for exports.

In total, imports require 645,000 tonnes of total catch while exports need 71, 000 tonnes (9 times less). Of this, the imported fish meal production required 597,000 tonnes or 92 per cent.

The requirements from the marine ecosystems, expressed as sea surface area needed to produce the harvested fish can be estimated by using the ratio between fish biomass and primary production. The area needed for the solar energy fixation was in the order of 171,000 km^2 for imports and 20,700 km^2 for exports (Table 2).

Inputs of auxiliary energy used in fisheries and fish processing are shown in Table 2, expressed in solar energy equivalents. The estimated inputs are largest in fisheries, representing 73-80 per cent of the total auxiliary energy from the economy.

The direct energy output of the food produced is sometimes compared to the inputs of fossil energy (Hall et al., 1986). Here, the energy input in fisheries and processing is shown to be around 33 times the output in terms of energy content of the fish products for direct human consumption, and 9-33 times the energy content for fish meal, depending on the fishing method (purse seines use only 3 GJ per tonne while trawling requires an input of 24 GJ per tonne).

Table 3 compares the trade balance in monetary terms with the estimated ecological balance in solar energy terms. Both balances show an import surplus. However, while the import surplus is found to be 2.5 in terms of money, it is 7.1 in terms of embodied energy drawn from the marine ecosystems and the economic sectors. Thus, for every Swedish crown (SEK) paid for imported fish products, 1.4 GJ of embodied energy is required, while in exports each Swedish crown corresponds to 0.47 GJ. Of this the energy from the ecosystems is 1.3 and 0.07 GJ/SEK respectively (85 - 93 per cent).

Taking into account the type of fish product that is produced (fish for direct human consumption and fish for meal), there are large differencies. The estimated energy requirements from the ecosystem for fish for direct human consumption are 0.3 GJ per SEK, while the estimated embodied solar energy in fish used in meal production is 2.8 GJ per SEK, which is 9 times as much.

Table 3. The ratio between exports and imports of fish products in terms of money paid, traded goods, harvested fish biomass, direct and indirect (embodied energy) in solar-energy equivalents and the sea-surface area needed to produce the required fish harvest.

	Imports/Exports
Money	2.5
Traded goods	2.8
Fish biomass harvest	9.1
Embodied energy	7.1
Sea-surface area	8.3

7. DISCUSSION

For a small country like Sweden, trading goods and services is of vital economic importance. Trade provides access to foreign resources and in doing so exploits distant ecosystems for resources we lack or that other countries can exploit more efficiently. In the ideal case, there is mutual benefit for both countries. However, as is illustrated in Figure 4, the monetary flows cover only a smaller part of the support needed from the ecosystems to sustain the trade. The increased scale of exploitation of natural resources and awareness of the limits of ecosystem production capacity has focused management interests on sustainability. How does trade affect the sustainability of renewable resources?

In this study of the Swedish fish trade, 1.4 TJ of embodied energy per SEK is required in imports while only 0.47 TJ per SEK were needed to support exports. The significance of this difference in the energy to money relation is expressed in the following way: Sweden is presently buying high quality products and work performed in the marine ecosystems of other countries at a lower price than it is selling its own fish resources to those countries. Hence, from a national resource perspective this balance of fish trade is advantageous for Sweden.

This relation takes into consideration the work of the indigenous ecosystems and most of the embodied energy (85-93 per cent) comes from work processes in the marine ecosystem which is assumed to be a zero cost in the economy.

Overall energy balances have been estimated for several countries concentrating

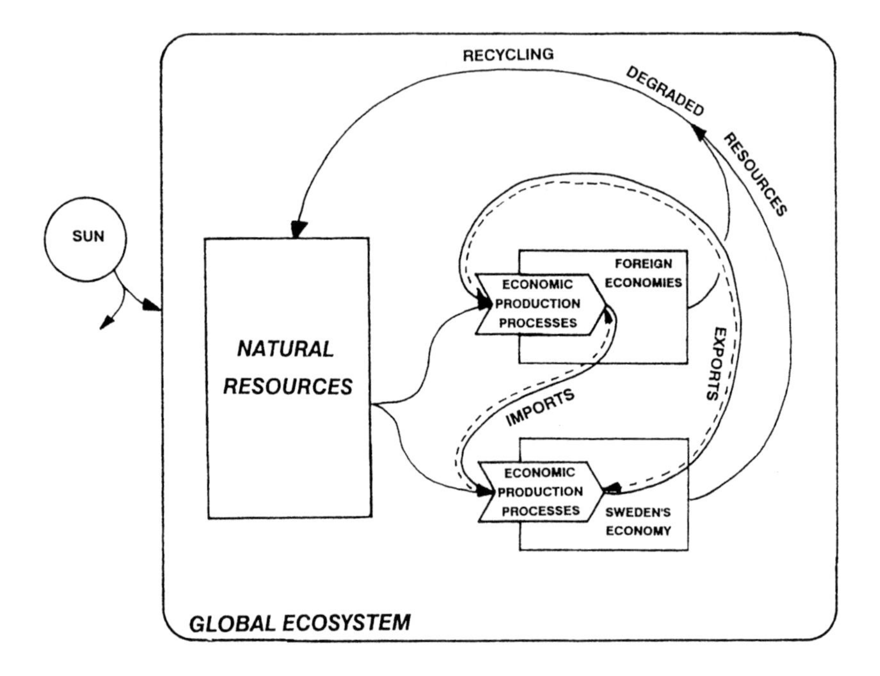

Figure 4. The inputs of natural resources to the economies as subsystems to the overall ecosystem, and the money flows to pay for goods and services between the economies.

on fossil energy use (Herendeen, 1978), but also in terms of embodied solar energy (Kemp et al., 1981; Odum & Odum, 1983). It has been found that there are often large differences between the embodied energy exported and that imported even though there is a balance of payment in terms of money (Odum, 1987). For example, New Zealand, which has an export structure directed towards raw products such as timber, meat and wool, exports more embodied solar energy than it receives even though foreign trade is fairly balanced in dollars (Odum & Odum, 1983; Odum, 1984). This is due to the fact that much of the necessary work performed in ecosystems, the ecosystem support, falls outside the framework of the economy and is not taken into consideration in economic evaluation.

The significance for the economy of this skewness appears to be minor as long as the scale of the economy relative to the environment is small since the regeneration of natural resources and processing of wastes by ecosystems are not limiting to economic activity (Daly, 1987; 1989). However, natural ecosystems sustain the extraction and exploitation of resources, process wastes, and recirculate materials and nutrients. These goods and services from the ecosystems provide at least as much to human welfare as man-made goods and services which are the only ones reflected in GNP (de Groot, 1987; Hueting, 1987; Odum, 1988). This study clearly illustrates this, since, for Swedish fish trade, the energy inputs from the marine ecosystems, an estimate of the ecosystem support, are 6 - 13 times larger than the estimated inputs of auxiliary energy used in the economy.

The imported fish products in my study require a total harvest of 645,000 tonnes of fish biomass. The ecosystem support area for this amount of fish is estimated to 171,000 km^2, while the corresponding support area for exports is 20,700 km^2. Renewable fish resources are thus being drawn from large ecosystem areas and concentrated in the Swedish economy. Sweden is therefore dependent on large life support areas in the sea to cover its present demand for fish protein. It is also likely that similarily large areas are needed for the assimilation and recirculation of waste products in the ecosystem (Folke et al., 1991).

The functional role of the fish in the ecosystem, providing feedback and support to other parts of the system, is altered or completely lost when humans, acting as a top predator, remove large amounts of fish biomass from the marine ecosystem (Jansson, 1988; Folke et al., 1991). Fisheries and fish farming are commonly treated as two separate industries without much regard to their overall efficiency or their dependence on the ecosystem resource base. However, fish farming depends heavily on the availability of fish protein as feed caught by fisheries. Sweden imports around 90 per cent of its fish meal demand and Swedish fish meal dependent industries (fish farming and livestock rearing) are relying on the underlying ecosystem support of this import. However, so far the dimensioning of these industries has been more or less completely decoupled from the production capacity of the supplying ecosystems. A large part of the fish meal imported to Sweden is supplied by Norway. Recently, the heavy fishing pressure on capelin, which is mainly used for fish meal production, severely reduced the capelin stocks in the Barents Sea and caused large decreases in predator populations such as the Norwegian cod stocks (Anon., 1987g). This overfishing of capelin created a wave of economic problems in Norwegian coastal fisheries and their development industries and communities (Folke & Kautsky, 1989).

When natural resource exploitation approaches the production limits of the ecosystems it becomes vital to recognize the interdependencies between ecological and economic systems in order to develop exploitation patterns that are both ecologically and economically sustainable.

On many fishing grounds, fishery exploitation has been balancing on the edge of overexploitation. It has been said that the major problem in fisheries management has shifted from finding and catching the fish to protecting fish stocks from overexploitation (Cushing, 1982; Gulland, 1985). At the same time, FAO has forecast the world demand for fish and shell fish products in the year 2000 to be 120 million tonnes as compared to 90 million tonnes in 1986 (Anon., 1989).

Exploitation of fish stocks has so far been concentrated on output and yield and the availability of low priced fossil fuels has facilitated increases in fishing effort (Rawitscher & Mayer, 1975). This postpones the economic incentive to reduce fishery when recruitment of stocks decreases and induces overfishing with fish population crashes as a consequence. Also, the output of energy per unit input (catch per effort) has declined considerably as has been shown as a general trend, e.g. for Swedish Gotland fishery (Jansson & Zuchetto, 1978) and U.S. fishery (Hall et al., 1986). The energy return in terms of catches is usually considerably lower than the input of auxiliary energy in modern fisheries. In this case study of Sweden's exports and imports, an input of 9-33 Joules of auxiliary energy per Joule of catch is in agreement with the values found for many sea fisheries (Hall et al., 1986).

Using fish for meal production has been considered an efficient way to exploit "excess" fish stocks in the sea, and for upgrading fish products by using fish meal as input in fish farming and livestock rearing. Scrap fish is harvested in large quantities and handled as a bulk product which lowers the economic cost, and fish meal is sold at daily world market prices (Ordelius, Lantmännens Riksförbund, pers. comm.). The available energy output from fish meal in relation to inputs of auxiliary energy in fisheries and processing is 0.03 if trawlers are used or 0.11 if large scale purse seine fishery is used. For fish for human consumption the corresponding figure averages 0.03. Since fish meal is used as an input in fish farming and livestock rearing, additional auxiliary energy inputs are needed before the natural resource is transformed into the food demanded by human consumers. However, the contributions from ecosystem life-support systems required to produce the fish, e.g. herring, are as high irrespective of whether the fish is used for meal production or for direct human consumption. According to the present study, of the amount of energy drawn from the ecosystem, it takes 2.8 GJ of solar energy to produce one crown's worth of fish for meal, while for fish for direct human consumption the corresponding figure is 0.3 GJ. This means that similar ecosystem resources are estimated quite differently depending on how they enter into the economy.

If Sweden, for example, wishes to become self-sufficient in fish meal and substitute imports with domestic production based on Swedish fishery. This would roughly mean an additional 600,000 tonnes of fish harvest, based on the 1986 demand. This can be compared to the total allowable catch quotas (TAC), set for the entire Baltic Sea in 1986, of 480,000 tonnes for herring and sprat combined.

A sustainable exploitation of living aquatic resources will need to shift the emphasis towards qualitative aspects rather than solely on increasing quantity. As stated by Daly (1989): "what is being sustained in sustainable development is a level, not a rate of growth, of physical resource use and what is being developed is the qualitative capacity to convert that constant level of physical resource use into improved services for satisfying human wants." In order to meet human demand for high quality food products from the marine systems in a sustainable way, emphasis needs to be changed from maximizing output towards increasing input efficiency, where the ecological functions are included and consideration is taken to the physical constraints set on exploitation of marine ecosystem by production and carrying capacity (Cleveland, 1987; Jansson et al., 1987; Odum, 1989).

8. CONCLUSION

The possibility of exchanging resources between different areas is of importance for an efficient use of unevenly dispersed natural resources. Sweden as a net importer of fish products, relies on large ecosystem support areas for its consumption of fish proteins. Using biophysical estimates of this export/import relation shows that Sweden is much more dependent on ecological support than is indicated in standard economic evaluation. As exploitation of renewable resources approaches the capacity limits of ecosystem production, recognition of these ecosystem support requirements for producing and processing these resources becomes crucial.

ACKNOWLEDGEMENT

I would like to express my deepest gratitude to AnnMari Jansson for all her support, interest and help during my work with this article. Bob Herendeen and Cutler Cleveland provided constructive comments on the manuscript. Carl Folke gave me valuable help and encouraged me. This study was funded by The Bank of Sweden Tercentenary Foundation.

NOTES

1. *Biophysical* refers to the flows of energy and matter in the ecosystems and the economy which are necessary for the production of goods and services and functioning of human settlements (Cleveland, 1987; Jansson, 1989).
2. The *energy quality* concept. All forms of energy do not have the same ability to accomplish work per unit energy (e.g. Joule). More concentrated forms of energy, like oil, have a higher work potential than diluted forms such as sunlight, and oil is said to be of higher energy quality.
3. An alternative method that would give a somewhat lower primary production is to trace and follow the energy flows in the food web and to estimate transfer efficiencies between different trophic levels (Jansson & Zuchetto, 1978). However, there are difficulties in including especially the indirect flows.
4. The *energy cost of labor* has been estimated in a number of ways ranging from the direct energy required for the workers metabolism to the total energy used in society to support the workers standard of living (see e.g. Odum, 1971; Jansson & Zuchetto, 1978; Hall et al., 1986).

REFERENCES

Anonymous. 1984. Bättre yrkesfiske. Betänkande av utredningen av fiskefrågor. Ds Jo 1984:6. Jordbruksdepartementet, Stockholm.

Anonymous. 1985. Udnyttelse af organisk affald fra fiskeindustrien till energiframstilling. Fase 1: Energipotentiale. Vandkvalitetsinstituttet ATV, Hörsholm.

Anonymous. 1986a. Fiskeforkatalog 1986/1987. FK-EWOS, Södertälje.

Anonymous. 1986b, c, 1987a. Statistiska Meddelanden J 55 SM 8601: J54 SM 8601: J 55 SM 8701. Statistics Sweden, Stockholm.

Anonymous. 1987b. Import och export av fisk och fiskprodukter jan-dec 1986 och 1985. Statens Jordbruksnämnd, Jönköping.

Anonymous. 1987c. Kort om svenskt yrkesfiske 1986. Fiskeristyrelsen, Gothenburg.

Anonymous. 1987d. Tabeller över produktion och förbrukning av fodermedel. Juli 1985-juni 1986. Statens Jordbruksnämnd, Jönköping.

Anonymous. 1987e. Fishery Statistics. 1985. Statistics Norway, Oslo.

Anonymous. 1987f. Statistical Yearbook, vol. 91. Statistics Denmark, Copenhagen.

Anonymous. 1987g. Naturresurser og Miljö 1987. Report 88/1. Statistics Norway, Oslo.

Anonymous. 1988. *Yrkesfiskaren* nr 4/88, p. 10, and 4/89 p. 10.

Anonymous. 1989. Svensk Havsresursverksamhet på 90-talet. Förslag till Övergripande program. Delegationen för Samordning av Havsresursverksamheten. Stockholm

Boynton, W.R., Kemp, W.M., Hermann, A.J., Kahn, J.R., Schueler, T.R., Bollinger, S., Lonergan, S.C., Stevenson, J.C., Twilley, R., Staver, K. and Zucchetto, J.J. 1984. An Analysis of Energetic and Economic Values associated with the Decline of Submerged Macrophytic Communities in Chesapeake Bay. In: Mitsch, W.J. and Bosserman, R.W. (eds.). Energy and Ecological Modelling. Elsevier, Amsterdam. pp. 441.454.

Cleveland, C.J. 1987. Biophysical Economics: Historical perspective and current research trends. *Ecological Modelling* 38:47-73.

Cleveland, C.J., Costanza, R., Hall, C.A.S. and Kaufmann, R. 1984. Energy and the U.S. Economy: A Biophysical Perspective. *Science* 225:890-897.

Costanza, R. 1980. Embodied Energy and Economic Valuation. *Science* 210:1219-1224.

Cushing, D.H. 1982. Climate and Fisheries. Academic Press Inc., London.

Daly, H. 1987. The Economic growth Debate: What Some Economists Have Learned But Many Have Not. *Journal of Environmental Economics and Management* 14:1-14.

Daly, H. 1989. Sustainable Development: Some Basic Principles. Environmental Division, The World Bank, Washington (mimeographed).

de Groot, R. S. 1987. Environmental Functions as a Unifying Concept for Ecology and Economics. *The Environmentalist* 7:105-109.

Elmgren, R. 1984. Trophic Dynamics in the Enclosed, Brackish Baltic Sea. *Rapports et Proces-verbaux des Reunions, Conseil International pour l'Exploration de la Mer* 183:152-169.

Folke, C. 1988. Energy Economy of Salmon Aquaculture in the Baltic Sea. *Environmental Management* 12:525-537.

Folke. C, and Kautsky, N. 1989. The Role of Ecosystems for a Sustainable Development of Aquaculture. *Ambio* 18:234-243.

Folke, C., Hammer, M. and Jansson, A.M. 1990. The Life-Support Value of Ecosystems: A Case Study of the Baltic Sea Region. *Ecological Economics* 3:in press.

Gulland, J.A. 1984. Epilogue. In: May, R. M. (ed.). Exploitation of Marine Communities: Dahlem Konferenzen. Life Sciences Report 32. Springer Verlag, Berlin. pp. 335-338.

Hall, C.A.S., Cleveland, C.J. and Kaufmann, R. 1986. Energy and Resource Quality: The Ecology of the Economic Process. John Wiley and Sons, New York.

Hammer, M. 1990. Ecological Aspects of the Baltic Fisheries, and Relations to the Economy. *Contributions from the Askö Laboratory* No 36, Stockholm University, Stockholm.

Harden-Jones, F.R. 1974. (ed.). Sea Fisheries Research. Elek Science. London.

Herendeen, R. 1978. Energy Balance of Trade in Norway, 1973. *Energy Systems and Policy* 2:425-431.

Herendeen, R. and Bullard, C.W. 1974. Energy Costs of Goods and Services, 1963 and 1967. Document No. 140. Center for Advanced Computation, University of Illinois, Urbana, Illinois.

Hueting, R. 1987. An Economic Scenario that gives Top Priority to Saving the Environment. *Ecological Modelling* 38: 123-140.

Jansson, A.M. 1988. The Ecological Economics of Sustainable Development: Environmental Conservation Reconsidered. In: Stockholm Group for Studies on Resources Management (eds.). Perspectives of Sustainable Development: Some Critical Issues Related to the Brundtland Report. Stockholm Studies in Natural Resources Management No 1. Askö Laboratory, Division of Natural Resources Management, Stockholm University, Stockholm.

Jansson, A.M. 1989. Energy in Ecosystems. In: Wieser, W. and Gnaiger, E. (eds.). Energy Transformations in Cells and Organisms. Proceedings of the 10th Conference of the European Society for Comparative Physiology and Biochemistry. George Thieme Verlag, Stuttgart. pp. 302-309.

Jansson A.M. and Zucchetto, J. 1978. Energy, Economic and Ecological Relationships for Gotland, Sweden: A Regional Systems Study. *Ecological Bulletins* No 20. Stockholm.

Jansson, A.M, and Jansson, B.-O. 1988. Energy Analysis Approach to Ecosystem Redevelopment in the Baltic Sea and the Great Lakes. *Ambio* 17:131-136.

Jansson, A.M., Folke, C. and Hammer, M. 1987. Economics of Renewable Marine Resources: Man as a Waste Producing Top Consumer in the Baltic Sea. Paper presented at the Vienna Centre Conference on Ecology and Economics, September 26-29, 1987, Barcelona, Spain.

Kemp, W.M., Boynton, W.R. and Limburg, K. 1981. The Influence of Natural Resources and Demographic Factors in the Economic Production of Nations. In: Mitsch, W.J. and Bosserman, R.W.(eds.). Energy and Ecological Modelling. Elsevier, Amsterdam. pp. 827-829

Lavine, M.J. and Butler, T.J. 1982. Use of Embodied Energy Values to Price Environmental Factors: Examining the Embodied Energy/Dollar Relationship. Cornell University, Ithaca, N.Y.

Leach, G. 1976. Energy and Food Production. IPC Science and Technology Press, Guildford.

Limburg, K. 1986. Gotland's Fisheries. A Case Study of the Ecolonomic/Ecological Processes of Renewable Resource Exploitation. *Contributions from the Askö Laboratory* No 31. Stockholm University, Stockholm.

Lorentzon, G. 1978. Energibalansen i den norske fiskerinaering. Norges Fiskeriforskningsråd, Oslo (mimeographed).

Malmberg, S.-A. 1988. Ecological Impact of Hydrographic Conditions in Icelandic Waters. In: Wyatt, T. and Larraneta, M.G. (eds.). Long Term Changes in Marine Fish Populations. Instituto de Investigaciones Marinas de Vigo, Vigo. pp. 95-123.

May, R. M. (ed.). 1984. Exploitation of Marine Communities: Dahlem Konferenzen. Life Sciences Report 32. Springer Verlag, Berlin.

Nixon, S.W., Oviatt, C.A., Frithsen, J. and Sullivan, B. 1986. Nutrients and the Productivity of Estuarine and Coastal Marine Ecosystems. *Journal Limnological Society of South Africa* 12:43 -71.

Odum, E.P. 1989. Input Management of Production Systems. *Science* 243:177-182.

Odum, H.T. 1983. Systems Ecology. John Wiley and Sons, New York.

Odum, H.T. 1984. Embodied Energy, Foreign Trade and Welfare of Nations. In: Jansson, A.M. (ed.). Integration of Economy and Ecology: An Outlook for the Eighties. Proceedings from the Wallenberg Symposia. Askö Laboratory, Stockholm University, Stockholm. pp.185-199.

Odum, H.T. 1987. Models for National, International and Global Systems Policy. In: Braat, L.C. and Van Lierop, W.F.J. (eds.). Economic Ecological Modeling. North-Holland, Amsterdam. pp. 203-251.

Odum, H.T. 1988. Self-Organization, Transformity and Information. *Science* 242: 1132-1139.

Odum, H.T. and Odum, E.C. 1983. Energy Analysis Overview of Nations. Working Paper. International Institute for Applied Systems Analysis, Laxenburg.

Pimentel, D. and Pimentel, M. 1979. Food, Energy and Society. John Wiley & Sons New York.

Pitcher, T.J. 1977. An Energy Budget for a Rainbow Trout Farm. *Environmental Conservation* 4:59-65.

Rawitscher, M. and Mayer, J. 1977. Nutritional Output and Energy Inputs in Seafoods. *Science* 198:261-264.

Steele, J.H. 1974. The Structure of Marine Ecosystems. Blackwell, Oxford.

Watanabe, H. and Okubo, M. 1989. Energy Input in Marine Fisheries of Japan. *Nippon Suisan Gakkaishi* 55:25-33.

Part III

ENVIRONMENT AND ECONOMY IN DEVELOPING COUNTRIES

Linking the Natural Environment and the Economy;
Essays from the Eco-Eco Group,
Carl Folke and Tomas Kåberger (editors)
Second Edition.
1992. Kluwer Academic Publishers

CHAPTER 11

Use and Impacts of Chemical Pesticides in Smallholder Agriculture in the Central Kenya Highlands

by

Carl Christiansson

Department of Ecology and Environmental Research
Swedish University of Agricultural Sciences
Box 7013, S-750 07, Uppsala
and
Department of Physical Geography
University of Stockholm
S-106 91 Stockholm, Sweden

During the period 1977 - 88 trends in smallholder agriculture in central Kenya were studied. One of the trends observed was a rapid increase in the use of agricultural pesticides. This text presents the results of surveys of pesticide use carried out in 1982 - 86 in two locations in the Central Kenya Highlands. An account is given of marketing and use of pesticides as well as an insight into the ways in which the farmers get information about how to handle toxic chemicals. Representatives of 140 households were interviewed using standardized questionnaires. The data thus collected were supplemented by semistructured interviews with Government officials, salesmen, shopkeepers, etc. The study shows that agricultural pesticides are widely used in smallholder agriculture in the Kenya Highlands. Considerably more pesticides are used on cash crops than on subsistence crops. Pesticides are more frequently used in areas where coffee is the primary cash crop compared to areas where potatoes or vegetables are produced for cash purposes. Many pesticides not allowed in industrialized countries due to their negative effects on the human health and on the environment are widely used in the study area. The farmers' knowledge is limited concerning both the correct handling of the products and personal health protection. The reason is a severe lack of information from the authorities, the pesticide manufacturers and dealers. A factor that limits the use of suitable protective equipment is the cost of such items. More than 50 per cent of the farmers interviewed in 1986 had experienced health problems related to what in many instances appeared to be the handling of the pesticides.

1. INTRODUCTION

In 1977, the present author, together with colleagues at the Departments of Physical and Human Geography, University of Stockholm, initiated an integrated study of some smallholder areas in the Central Kenya Highlands. The study focused on problems concerning the natural environment, the land use and the socio-economic conditions. Field data were collected during a period of twelve years (1977 - 1988). Most of the data collection took place during annual field courses run by the Geography Departments of Stockholm University. During the courses the Swedish participants worked together with Kenyan field assistants recruited among local school teachers and secondary school leavers.

The study covers more than 3,800 individuals. Most of them live on the 350 smallholdings which, within the scope of the project, have been studied in detail and mapped with respect to soil conditions, land use, soil conservation, buildings, etc. The results have been published in fifteen volumes in the series Geographical Studies in Highland Kenya (Departments of Physical and Human Geography, University of Stockholm, 1977-88). Of the total number of households investigated in the main project 140 were included in the pesticide study presented here.

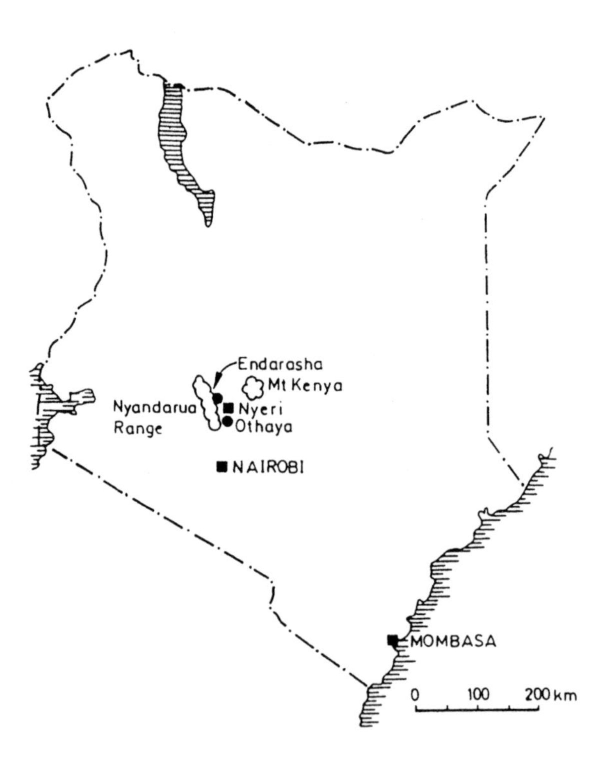

Figure 1. The study areas are situated on the eastern slopes of Nyandarua Range in the central Kenya highlands.

One of the study areas, Endarasha (2,500 metres above sea level) situated 30 km north of Nyeri, the provincial capital of Kenya's Central Province, is a so-called "settlement scheme", i. e. earlier European owned land that has been divided into several hundred smallholdings of 3 - 10 hectares. The most important subsistence crops in Endarasha are maize, beans, vegetables and root crops, particularly Irish potatoes. Potatoes, vegetables and pyrethrum also serve as cash crops. Cattle for milk production play an important role in the local economy.

The other area in the pesticide survey, Othaya (1,800 metres above sea level), is situated 20 km south of Nyeri (Figure 1). It is a farming area which during the colonial period had the status of an African reserve. Through a land reform in the 1950's and 60's, the farms which were in scattered fields were assembled into consolidated units. Most of the farms investigated are situated in the village of Gatugi in Thuti sublocation 3 km east of Othaya township, the administrative and commercial centre of the Othaya division.

The size of the farms in Thuti ranges between 1 and 5 hectares. Due to the lower altitude the variety of crops grown is somewhat greater than in Endarasha. The most important subsistence crops are maize, beans, sweet potatoes, Irish potatoes, bananas and vegetables. The most important cash crop in Thuti is coffee. Livestock is kept but is of less importance to the local economy than in Endarasha.

One of the tendencies observed in the study areas during the project was the increased use of agricultural chemicals, particularly pesticides. For the farmers this has both agronomic and economic implications as well as consequences for the health situation. Adverse effects on the environment have also been observed.

2. OBJECTIVES AND METHODS

The aim of the present text is to provide insight into the availability and use of agricultural pesticides in two smallholder areas in central Kenya. Insight is also given into the ways in which the farmers get information about the properties of the products and how to handle them. Further, the text aims to describe and discuss health impacts on the farming population and general impacts on the local environment. The data collection was based on traditional geographical field methods, partly of the type termed "Rapid Rural Appraisal" (RRA) (McCracken et al., 1988). However, the term had not yet been introduced in scientific contexts when our studies were initiated.

By means of a standard questionnaire, representatives of 140 rural households were interviewed on the use and effects of pesticides employed in small scale crop production. Those data were supplemented by more detailed interviews with selected farmers, with the administrative and field staff of the Ministry of Agriculture, health workers, salesmen and shopkeepers, coffee factory staff and a representative of the Environmental Liaison Centre in Nairobi. The results presented below are based on studies carried out during the period 1982 - 1986.

The statistical material is small and it has weaknesses of the kind which is common when data is collected by many interviewers. Moreover, the survey was never designed and structured once and for all. As field work continued over several years, the data collection had to be adapted to succesively changing conditions. However, the wording of the questions which form the basis of the discussion in the present text was the same from year to year. In some years, though, extra questions were added to throw light on a particular problem. When it was considered that we knew enough about the problem, the question was dropped.

The results indicate trends in pesticide use and attitudes to toxic agricultural chemicals in the study areas. In physical terms the results are representative of the conditions on a small percentage of Kenya's land area. However, this area houses a substantial part of the population of the country. Still, no far-reaching conclusions will be drawn since that would require a larger sample population and a different type of study.

3. WHAT IS A PESTICIDE?

A pesticide can be defined as a chemical substance used for protection against damage or against sanitary or other inconveniences caused by biological organisms (for instance animals, weeds, fungi, bacteria or virus) (Singh, 1983). Chemical pesticides are sub-divided, depending on their target, into *insecticides, herbicides, fungicides, bactericides* and *rodenticides*.

4. THE USE OF PESTICIDES IN DEVELOPING COUNTRIES

At present, pests, including fungi, vermin, insects, etc. destroy well over 1/3 of the annual world production of food. In temperate areas losses may reach 20-30 per cent of the potential production, while for instance in Southern Asia and Africa the losses are well over 40 per cent (Singh, 1983; Ekström & Hofsten , 1985). In Bangladesh 15 per cent of the annual production of food grains is lost to pests (Ahmad, 1987). For an important cash crop like cotton, the losses during 1980 in the whole of Africa amounted to 41 per cent. The figure was far higher for several individual countries, for instance 53 per cent in Uganda and Chad (Woller, 1983). About 30 per cent of Africa's food is destroyed in storage alone (Singh, 1983). These facts form the basis for an expanding market for agricultural chemicals.

During the 1970s' the use of pesticides worldwide increased by 4-5 percent per year (Berglund & Johansson, 1982). From time to time between 1960 and the early

1980's the annual increase was as high as 15 per cent or more (Bull, 1982).

From 1970 to 1980 the real value of pesticide imports to the third world increased more than sixfold (Forget, 1989). The cost of pesticide imports to the Gezira cotton project in Sudan increased by 1,400 per cent between 1966 and 1981 (Bull & Myers, 1984). In Kenya 625 million Ksh were allocated to imports of pesticides in 1988. That represents a 50 per cent increase over the 1984 figure (Orlale, 1989).

This trend is well illustrated in the study areas where, according to shopkeepers interviewed, the sale of agricultural chemicals has increased considerably during the 1980's. This increase is due to the fact that the production of cash crops for export, such as coffee, has increased within the smallholder sector. Much larger quantities of pesticides are used on cash crops than on crops grown for subsistence (Weir & Schapiro, 1983; Ekberg et al., 1985). In the study presented here, coffee is clearly the most frequently sprayed of all crops.

With "the green revolution" high yielding hybrids of maize and rice were introduced in many developing countries. This has to some extent improved the future prospects for the world's food supplies. But these hybrids require more fertilizers than traditional breeds, and they are much more sensitive to attacks by insects and diseases. This leads to a situation where the farmer is forced to use more pesticides (Forget, 1989). Overuse of pesticides also occurs as a direct consequence of aggressive marketing (Bull & Meyers, 1984). Ahmad (1987) reports from Bangladesh how campaigns by multinational companies selling pesticides in the country have succeeded in linking the use of chemical pesticides with modernism in the minds of the farmers. The result is that there is little interest in alternative means of pest control.

5. PESTICIDES IN RURAL KENYA

5.1 Marketing and Information

From advertisment boards in industrialized countries to the walls of rural homes in Kenya, pesticide advertisments are part of the scene. The message is the same: You need our pesticide for a successful harvest. This was noticed many times during our study. Large posters and colourful calendar pictures advertising agricultural chemicals are the most common decorations in the offices of government officers and in the farmers' homes.

The common attitude towards agricultural chemicals is open in Kenya, and there are no difficulties in acquiring information about pesticides or fertilizers. There are, however, restrictions about sales statistics on the national level for commercial reasons.

Chemical pesticides are produced, marketed and sold in Kenya by a number of different companies: Murphy, May and Baker, Kleenway, Crop Protection Chemicals, Bayer, Shell, etc. Most of these are branches of 24 transnational companies that control 80 per cent of the world market for pesticides (Bull & Myers, 1984).

When a new pesticide is introduced, the producing company distributes leaflets to the retailers, and after a month or so the representative visits the shops to give further information about the product. In general, representatives of the chemical firms regurlarly visit the shops that sell pesticides, to make stock inventories. When any item they sell is out of stock this is pointed out to the shopkeeper, so that he can order the missing product.

An important aspect of the companies' promotion is the arranging of conferences to which the retailers are invited. These conferences are usually held at the best hotels in the area. The participants are offered free board and lodging while the companies inform them about their products. The most active participants receive gifts, often in cash, at the end of the conference.

The chemical firms also arrange field demonstrations on selected farms. The owners get their crops sprayed and are allowed to keep what remains of the product used. Pesticide information is also communicated to the farmers by representatives of Kenya Farmers Association (KFA) and by the Agricultural Department's local staff.

When interviewed, the owner of one of the privately owned shops in Gatugi (Thuti) stated that he gets all the information he feels is necessary from the producer/wholesale companies. The representatives of the companies from which he buys pesticides visit him approximately twice a month, first to deliver the pesticides and the second time to collect money for products sold. When new products are introduced or when existing products are changed, he is advised to inform the farmers about how the doses should be adjusted.

The radio programme "Voice of the Farmer" - broadcast once a week - also gives information to farmers. The official link between the Government, the chemical industry and the farmers is the Chief Pesticide Officer at the Ministry of Agriculture.

5.2 Pesticides Available and Crops Sprayed

In Kenya the majority of the population are engaged in agriculture. They are directly dependent on reasonable harvests for their survival and cash income. The yields can, at least in the short run, be increased by the use of pesticides. The pesticides are expensive, but all farmers who can afford them buy these products.

Table 1 shows the use of some common pesticides on 65 farms investigated in Thuti and Endarasha. Some farmers use two or more kinds of pesticides on, for instance, coffee which is the most heavily sprayed crop, followed by vegetables and root crops. In a study on 13 farms in Thuti in 1985 it was shown that more than 90 per cent of the total amount of pesticides was applied on coffee. This agrees well with the findings from studies carried out by other researchers (Orlale, 1989).

Table 1. Use of Agricultural Pesticides in Thuti and Endarasha, April 1984 and 1986.

The figures show how many of 65 farmers interviewed who use the respective product, and on which crops. Copper, Dithane M45, Malathion and DDT are the most used pesticides. Coffee is the crop on which pesticides are most commonly used. Coffee is not grown in Endarasha, however, but there, pesticides are frequently used on root crops, vegetables, beans and peas.

	Pesticide							
Crop	**Copper red/ green**	**DDT**	**Dithane M45**	**Mala-thion**	**Difo-latan**	**Ambush**	**Others[a]**	**Un-known**
Grain crops	0	10	1	0	0	1	8	2
Beans/Peas	1	5	8	0	0	1	6	3
Potatoes/ Root crops	13	3	17	0	0	1	8	1
Vegetables	2	6	11	0	0	1	13	2
Coffee	40	0	0	22	6	0	8	1
Fruit trees	4	0	0	1	0	0	2	0

Notes to Table 1.

a. Actellic-l, Aldrin, Dieldrin, Fenitrothion, Gusathion, Lebaycid Ridomil, Sumithion.

In Endarasha, DDT and Dithane M 45 were the most commonly used pesticides applied mainly on maize and potatoes according to surveys in 1983 and 1984. By 1986 DDT had lost in importance. It should be noted that coffee is not grown in Endarasha due to the cooler climate.

The water-soluble pesticides are applied by means of 15-20 litre hand-pumped knapsack sprayers. These can be bought or hired from the co-operative or private shops or from the coffee factories.

5.3 Some Common Crop Pests and Their Recommended Cures

Economically, the most important crops in the study areas are coffee, maize, potatoes and vegetables. Of these coffee is particularly sensitive to diseases and it is frequently attacked by "leaf rust" (Hemileia vastatrix) and "coffee berry disease" (CBD) (Colletotricum coffeanum)(Acland, 1971; Christiansson & Strand, 1983). Both these diseases are caused by fungi and they occur in the form of tinted berries and leaves. CBD almost brought about total destruction of the coffee industry in Kenya in its peak years, 1967-68, and in the early 1980's the combined loss from leaf rust and CBD still amounted to around 25 per cent of the potential harvest. Copper solution (green, blue or red) is active against both these pests. The dose recommended by the Agricultural

Department varies, depending on how large the risk is for attacks and the geographical position of the coffee district. According to the farmers interviewed in Thuti, coffee is sprayed by green, blue or red copper between five and eleven times per year. The most common pattern is for the coffee trees to be sprayed every three weeks after they have flowered until the berries are picked. However, not all farmers can afford to buy the amounts recommended by the advisors - the extension staff. Thus they tend to cut down on the number of sprayings or they use concentrations lower than those advised. This may result not only in fewer insects being killed but even in increased reproduction within the target group (Pimentel & Edwards, 1982) (cf. Figure 2). The spraying is car-

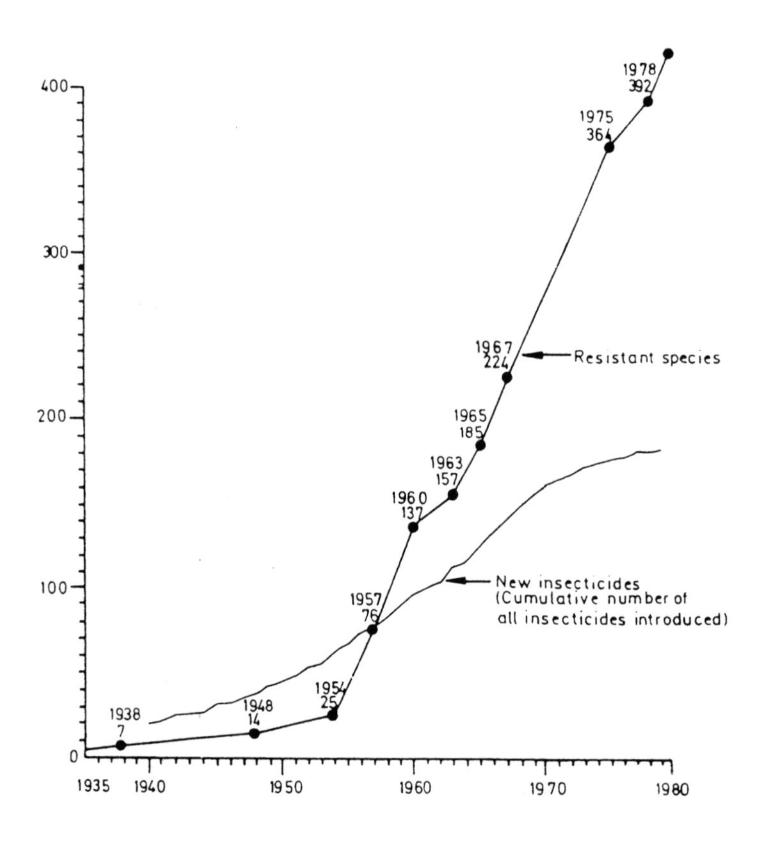

Figure 2. Resistant species of arthropodes and new insecticides 1938 - 1980. (After Bull, 1982).

In the mid-1960s the extent of the problem of reduced effects of pesticides on many pests became obvious. Because insects have a rapid regeneration cycle they soon build up resistance to different insecticides (Boza Barducci 1972; Bull & Myers 1984). In 1960, 137 species of arthropodes (insects, spiders, millipeds and crustaceans) were resistant to two or more kinds of insecticide. In 1980 the number had more than trebled to 432. Even among rats, fungi, bacteria and weeds, resistance has increased (Bull 1982; Pimentel & Edwards 1982).

ried out during the rainy seasons, March-May and October-November. In addition to copper, or as an alternative, Difolatan, Malathion and Dithane M 45 are used on coffee.

When the pesticide used is Difolatan, which is more efficient than copper, sprayings can be less frequent. However, at the time of our survey in the mid 1980s, Difolatan was five times as expensive as green copper.

An insect larva that sometimes affects the coffee trees is the "leaf miner" (Leucoptera sp.). At present the Agricultural Department recommends spraying against leaf miner only when large numbers of larvae are found on the trees. In the 1970's systematic spraying with DDT against leaf miner was common, however, this caused eradication of the natural enemies of the leaf miner (Ngugi et al., 1978). Another pest on coffee is "scales". Different varieties occur. According to Agriculture Department staff in Othaya, they are combatted with Malathion sprayed twice a year and/or Dieldrin 18 per cent placed around the stem. Moreover, the coffee fields are treated two or three times a year with herbicides such as Round Up or Gramoxone to reduce the weeds.

The coffee berries are carefully checked and classified before they are accepted at the factory. Thus it is impossible for the grower not to take part in the chemical campaign against diseases and destructive insects on the coffee trees.

A moth that in the larva stage attacks maize is the "stalk borer" (Busseola fusca). The agricultural extension workers recommend Dipterex against stalk borers, but this product is not used on any of the farms included in our survey. The reason for this is not known. However, several growers indicated they use DDT on maize and beans. This is also common practice in other maize-growing areas (Cox, 1985; Friis-Hansen, 1987). Beans are also attacked by "leaf lice", against which Thiodan is recommended by the extension staff. Cabbages are attacked by so-called "cut worms" (larvae of a moth) and against those the extension staff recommend Aldrin 2.5 per cent. Against insects in general on vegetables, Metasystox is suggested.

Amongst diseases affecting potatoes, an important cash crop in Endarasha and a common subsistence crop in both study areas, the fungus disease "early and late blight" (Phytophtora infestans). This disease is mainly treated by using Dithane M 45 or Ridomil which are applied two to four times during the growing season, but some farmers use copper on potatoes (Figure 3).

To protect them from insect attacks and fungi, tomatoes and cabbages are treated two or three times during the growth season of three months. For this purpose DDT, Ridomil and Dithane M 45 are commonly used.

The widespread use of pesticides in smallholder agriculture in the study areas not only keeps pests and harmful insect populations down, birds that control rodent populations and bees needed for pollination are also killed. Our informants in Othaya told us that the bird population had decreased sharply in recent years and monkeys were fewer than they used to be. From other parts of Kenya and from certain areas in Tanzania it is reported that bee-keeping has disappeared due to the intense use of insecticides (Bull, 1982). Where pesticides affect the soil fauna organic matter does not decompose properly and is not incorporated into the soil and this result in risk of degradation of the soil structure (Pimentel & Edwards, 1982). This in turn increases the erosion.

Figure 3. Spraying the potatoes on a farm in Endarasha. Drawing by H. Drake after a photo by L. Lundgren.

5.4 Trends in the Pesticide Sales Pattern

In the small township of Endarasha there are two shops that sell agricultural chemicals. In Othaya township and its immediate surroundings there are about ten shops which sell pesticides. Since 1983, several new shops selling agricultural chemicals have opened, indicating that these products are profitable.

The number of different products available has also increased markedly. One of the two general shops in Endarasha could offer its customers six different kinds of pesticides in April 1982. In 1983 this figure had increased to ten and by May 1984 fifteen different pesticides could be bought in the shop.

A survey in 1983 showed that there were 25 different pesticides for sale in five shops in Othaya. A year later, in 1984, there were 40 different kinds available in the shops investigated. In 1986, a total of 55 different products were found in seven shops in Nyeri and Othaya (Table 2).

Table 2. Pesticides Available in Seven Retail Shops in Nyeri and Othaya, April 1986

The following shops were included: Kareju Enterprise, Nyeri, Kenya Grain Growers Co-op Union Ltd, Nyeri, Nyeri District Co-op Union Ltd, Othaya, E M-D G Store, Othaya, Kenya Grain Growers Co-op Union Ltd, Othaya, W M Store, Gatugi.

Trade Name	**Active substance**	**Use**
Aldrin 2.5%, 5%	Aldrin (och)	Insecticide
Actellic-1	Pirimephosmethyl (oph)	-"-
Agrocide 3	Gamma BHC (och)	-"-
Aldrex 48%		
Afalon 50%	Linuron	Herbicide
Ambush	Permethrin	Insecticide
Antracol	Propineb	Fungicide
Ant Killex		
Baygon	Propoxur	Insecticide
Bayleton	Triadimefon	Fungicide
Birgin		
Bygolt	Flucytrinat	Insecticide
Dawa ya Mboga	Gamma BHC	-"-
Dedevap	Dichlorvos	-"-
Delnav DFF	(oph)	-"-
DDT 5%, 25%	DDT (och)	-"-
Diazinon	Diazinon (oph)	-"-
Dieldrin 18%	Dieldrin (och)	-"-
Didimac 5%	DDT (och)	-"-
Difolatan		Fungicide
Dimethoate	Dimethoate	-"-
Dipterex	Trichlorphon (oph)	-"-
Dithane M45	Mancozeb	-"-
Fenitrothion 50%	Fenitrothion	Insecticide
Follion Feed		
Gammexane	Gamma BHC	-"-
Gramoxone	Paraquat	Herbicide
Grain Store Spray		
Green Copper	Copper compound	Fungicide
Gusathion	Azunfozmetyl (oph)	Insecticide
Kilpest	Malathion (oph)	-"-
Kynadrin	Dieldrin (och)	-"-
Lebaycid	Fenthion (oph)	-"-
Lenasad		
Mancozam	Mancozeb	Fungicide
Malamite	Malathion	Insecticide
Malathion 2%, 50%	Malathion	-"-
Murphane	Zineb	Fungicide
Metasystox	Demeton (oph)	Insecticide
Nexion	Bromofos	-"-
Nogos	Dichlorvos	-"-

Table 2, continued

Pyrethrum	Pyrethroids	Insecticide
Racumin	Kumatetrayl	Rodenticide
Red Copper	Copper compound	Fungicide
Ridomil	Metalaxyl, Mancozeb	Fungicide
Round up	Glyphosat	Herbicide
Roxion	Dimethoat	Insecticide
Sevin 85	Carbaryl	-"-
Sumithion	Fenitrothion	-"-
Supadip		
Supona	Chlorfenvinphos	-"-
Thiodan 18%	Endosulphan	-"-
Tropical mortegg	Tar oil	-"-
Zelco Taste		
Zibycinade		

(oph) = organophosphate
(och) = organochlorine

The pesticides with the highest sales in the KFA shop in Othaya were Dithane M 45, Malathion and Copper. According to KFA's records, the demand for DDT has decreased lately. This may be due to the fact that in 1984 the debate on the environmental effects of DDT reached the study area, and it was no longer recommended by the extension staff as indiscriminately as before. By 1986 it could no longer be bought in all shops.

In the coffee factories in Othaya, the majority of chemicals sold are those used in coffee-growing. They are Copper, Dithane M 45, Difolatan, Malathion, Fenitrothion and earlier Gusathion. Many farmers use Malathion on coffee, but in the coffee factories in the study area they do not sell that product at all. Gusathion, which in 1983 was one of the products with the highest sales was no longer on the list of pesticides sold in 1984, possibly because it was never supplied by the producer (Christiansson & Strand, 1983; Christiansson & Wåhlin, 1984). The sale of Difolatan has decreased greatly since 1982 when false and ineffective products with that name appeared on the market. This illustrates the problems the farmers face in terms of the very large variety and the frequent changes in names and properties of products on the market (cf. Table 2).

The pesticide store in Gatugi sells products that are supplementary to those of the coffee factories. Among others a Diazinon product is available that according to the label is recommended against bed bugs. This product, which is also for sale in the store in Endarasha, is classified by WHO as "highly hazardous", which means that it is very toxic and if used in an industrialized country must only be handled by persons with a special permit.

It should be noted that the sale statistics for pesticides not only reflect the demand but also the supply of the products. The distribution of goods is irregular, and thus one or more of the products are often missing in the shops. It is therefore difficult to judge the popularity of a product based on the sales statistics. Sometimes the pesticides with the highest sales are those that are available, when in fact others are urgently needed.

Eighty per cent of the interviewed farmers in Thuti (in 1986) did not use what they considered to be the most efficient pesticides. These, many said, were too expensive, and moreover the shops do not always have the best products in stock.

The difference in price between practically identical products is sometimes considerable. As an example a small survey of prices in five shops in Othaya in 1983 revealed differences in the cost of the insecticide Malathion by more than 55 per cent, while for instance the herbicide Gramoxone was the same in all five shops (Christiansson & Strand, 1983).

6. PROTECTION OF PEOPLE HANDLING THE PESTICIDES

6.1 Protective Clothing

The protective measures in connection with spraying in the study areas are unsatisfactory. The principal reason why so few people use protective clothing when spraying is the widespread ignorance of the risks involved in handling the chemicals. This in turn seems to be due to insufficient information being supplied to the farmers by dealers and extension staff.

Another reason for the failure of farmers to use protective clothing is the high cost of this equipment. Complete safety equipment would consist of the following: protective goggles, respiratory filter, raincoat, rubber gloves, hat or cap and rubber boots. Such an outfit costs the equivalent of around US$ 100. Seven shops were checked, but only three sold the proper kind of equipment. According to the dealers, the demand for this kind of goods is minimal. The most commonly used protective item is a hat or a cap. In two cases in the present study this was supplemented by a raincoat. Thus, 11 out of 39 farmers interviewed use a hat as their only safety measure when spraying. Protective goggles were used by only two persons, while none used a respiratory filter.

Seventeen out of 39 interviewed farmers did not use any kind of protective clothing. Twenty-five per cent of those interviewed considerd the cost of the equipment to be too high, a further 25 per cent considerd the pesticides to be so harmless that no protective equipment was needed. Four farmers stated that they had now become used to the chemicals and no longer had any problems (Table 3). During our field work in Thuti and Endarasha we often saw children, 10-12 years old, wearing only shorts and a T-shirt or a sweater, assisting parents in the spraying of crops (Figure 4).

Table 3. Protective Measures taken when Spraying.

The sample population consisted of 39 adult persons engaged in farming in Thuti and Endarasha. Some of those interviewed mentioned two or more protective measures that they applied.

Protective measure	**Number of farmers applying the respective measure**
None	17
Gloves	3
Goggles	2
Hat	13
Coat/Raincoat	8
Leather jacket	2
Empty bag used as coat	4
Rubber boots	3
Unguent on face before spraying	1
Avoid talking when spraying	1
Handkerchief in front of mouth	2

7. Health Effects

Nearly 85 per cent of the world pesticide production is consumed in industrialized countries. Yet the incidence of recorded pesticide related health problems is 13 times higher in the Third World (Forges, 1989). Many chemicals used in agriculture give rise to allergic reactions, eczema, problems with mucous membranes etc. Some are suspected of causing cancer (McCracken & Conway, 1987; Orlale, 1989). Forges (1989) mentions a figure of close to two million people annually and worldwide being poisoned by pesticides and possibly up to 40,000 of them die as a result. There is a currently accepted figure that some 10,000 farmers in developing countries are killed every year due to accidental insecticide poisoning. However, it has been suggested by Loewinson, cited by McCracken and Conway (1987), that this figure is grossly underestimated. Loewinson studied the relations between mortality patterns and the use of pesticides in rice growing areas in the Philippines.

To minimize the negative effects of agricultural chemicals the users must be well informed about how to mix and apply them; they must know the proper doses, the relevant safety intervals, and they must be familiar with the published safety information. To gain this knowledge, it is essential to be able to read. In many developing count-

Figure 4. Children assisting in mixing chemicals before spraying the coffee. Drawing by H. Drake after a photo by the author.

ries the majority of the population are illiterate. At the time of the 1979 census 58 per cent of the population of Kenya was illiterate. Today the figure is around 48 per cent. However, among the sample population in the 1986 survey in Thuti and in Endarasha, 266 individuals above six years of age, only 15 per cent did not have primary education.

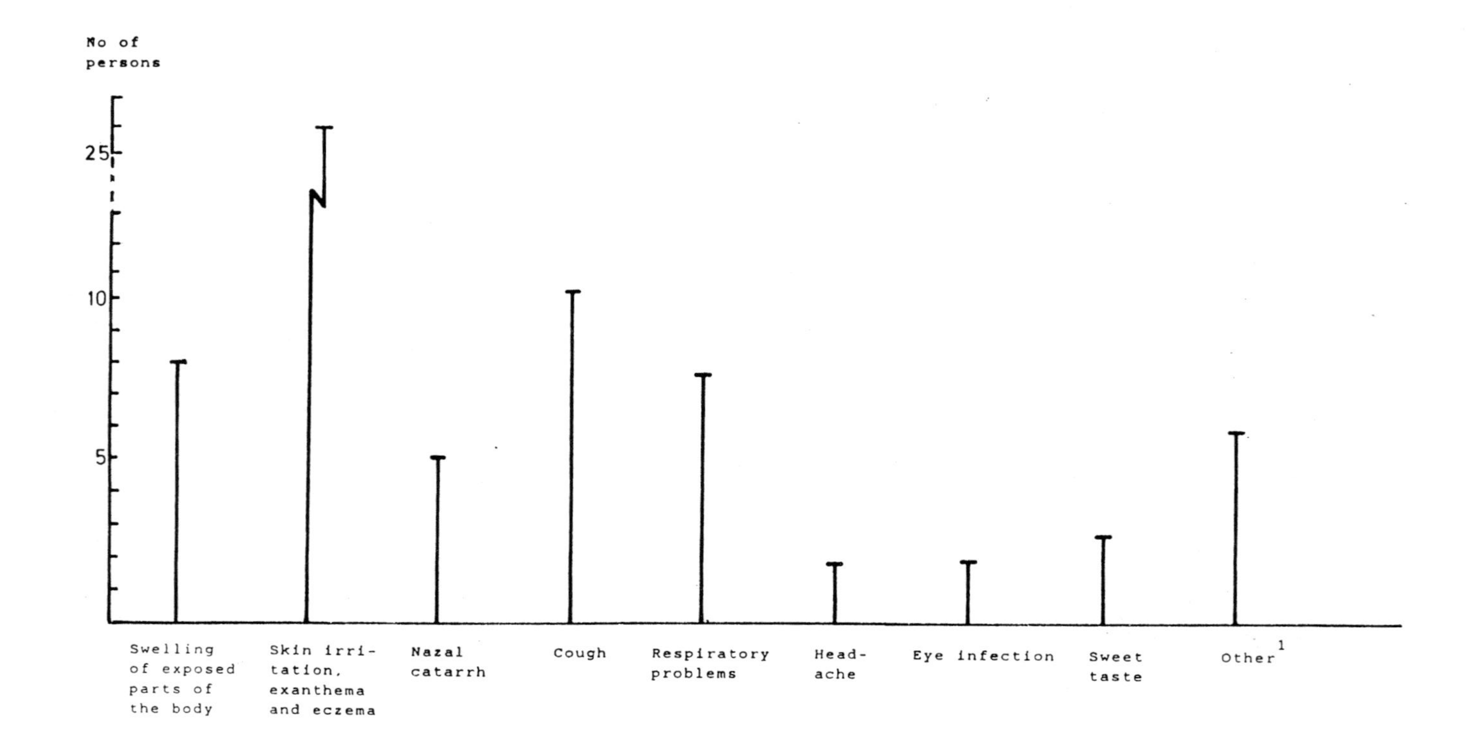

Figure 5. Health problems perceived by the interviewed as related to the use of pesticides. The sample population consisted of 78 adult persons engaged in farming in Thuti and Endarasha 1984-1986. Some of the persons interviewed mentioned two or more health problems that they had experienced (The sweet taste was caused in all cases by Copper Nordox. After spraying with this product, food, drinks, and cigarettes get a sweet taste for approximately 24 hours). It should be noted that some frequently used pesticides of the organo-phosphate group are known to attack the nervous system. However, no such problems were mentioned by the farmers interviewed.

In many developing countries pesticides can be bought in any empty can or soda bottle. Later the vessel may be used for drinks or as a toy for children. Orlale (1989) reports about such cases from a study by Mwanthi and Kimani in an area north of Nairobi.

On some packages there is a printed warning that in case of accident a doctor should be sent for. However, in the developing countries there are few doctors, and counteractive drugs are seldom available.

The storage facilities for chemicals in one of the general shops in Endarasha are an obvious example of how different the Kenyan attitude to pesticides is compared to that prevailing in the industrialized world. Behind an unlocked latticed door 15 different toxic products are stored, three of which (Thiodan, Diazinon and Metasystox) would be classified by WHO as extremely hazardous. A few products with toxic active substances belong to this risk class, which means that the products must be stored in a locked space and only persons with special knowledge and holding a special temporary permit may buy it. DDT is also sold in the shop, as are Aldrin, Dieldrin and Captafol which are banned in many industrialized countries (Orlale, 1989).

During our interviews several farmers pointed out that they had been informed by the agricultural extension staff that they should be careful when using pesticides. However, the majority of the farmers indicated that they had received no particular information about how to handle and store pesticides or about possible health risks. Their only information had been the text on the packages.

The health hazard is also increased by post harvest treatment of crops. A common practice is to treat maize cobs or leafy vegetables with insecticides to enhance their appearance before they are sold in the market (Christiansson & Strand, 1983; Christiansson et al., 1986; Friis-Hansen, 1987; McCracken & Conway, 1987).

Sixty per cent of the farmers interviewed in 1986 in Thuti and Endarasha and 11 out of 13 interviewed in 1985 in Thuti had experienced health problems that *could* be related to the use of pesticides. The different kinds of problems are illustrated in Figure 5. Most of the problems experienced by the farmers have not been long-lasting. However, the long-term effects of frequent exposure to pesticides are not well known.

In a recent study by Mwanthi and Kimani in the Kiambu district of Kenya, the research team met many people at the local health centre "suffering from pesticide related ailments such as skin irritation, asthmatic attacks, hypertension, constant headaches and diarrhea. Coffee sprayers were particularly hard hit and the team branded them *the high risk group* because of the long hours they spend handling the chemicals" (Another study carried out by the Department of Physiology at the University of Nairobi indicated that after only two hours of chemical spraying, the average coffee sprayer shows evidence of a 20 per cent reduction in lung capacity. Some coffee sprayers work 30 to 40 hours per week (Orlale, 1989)).

In the light of the above it is surprising to meet the widespread ignorance among the local staff of the Ministries of Agriculture and Health illustrated by the following results of interviews.

7.1 Interviews with the Agricultural Extension Staff and Health Officers in Othaya and Doctors at Endarsha and Othaya Hospitals

The Agricultural Extension Officer in Othaya stated in 1986 that he had never heard of a case of illness caused by pesticides. "The people are used to these products and the side effects are not serious. If they were, the products would not be allowed for sale on the market." The most alarming aspect of pesticides, according to him, was that some persons had used such products to commit suicide.

The Health Officer in Othaya was also interviewed about health aspects of the use of agricultural chemicals. He believed that proper information and advice about pesticides is communicated to all farmers via the shops, the co-operative society and the Ministry of Agriculture. He also mentioned that he had never heard of anyone who had experienced any health problems in connection with spraying. He said that if cases had occurred he would have been informed by the hospital. He concluded the interview by saying that there are no really dangerous pesticides on the market today.

In 1983 the doctor in charge at the Endarasha hospital claimed that she had never observed any health effects caused by pesticides among her patients.

In April 1984, the Clinical Officer at the Othaya Hospital stated that troubles such as skin irritation, running nose and coughing were common among his patients, but that the doctor seldom knew the reasons for the trouble. When the patients come to the hospital they are never asked whether they have recently used chemicals or not. Moreover, the most intensive spraying period coincides with a period of cool and moist weather, and thus it was just as likely that the above symptoms had been caused by virus or bacteria just as by agricultural chemicals.

In April 1986, the Clinical Officer in charge at the Othaya Hospital was interviewed. He believed that some illnesses could possibly be caused by contact with pesticides. However, during the last two years only two cases definitely caused by pesticides had been recorded at the hospital. Both were cases of swelling and irritation on exposed parts of the body, i.e. mainly hands and faces. The doctor did not know of any cases where children had been poisoned by chemical products. "Every family knows how these products shall be stored to be out of reach of children."

The doctor told us that he was not giving any direct information to the farmers about risks involved in handling of chemical pesticides. For that he relies on the agricultural extension staff and the dealers. He is, however, in contact with the local Officer of the Ministry of Health and reports to him in case anything unusual occurs. It had not been considered necessary to report the two cases just mentioned.

The doctor pointed out the difficulties in diagnosing health problems caused by agricultural chemicals. Usually a long time passes from the day the farmer falls ill until he visits the hospital. The farmers seldom give themselves time to visit the hospital, as the queues are very long. The doctor thought that many farmers ignore symptoms of illness because of the economic loss and loss of time involved in staying away from the farm.

8. LEGAL ASPECTS

During our study it became clear that neither the Agricultural Officer in Othaya nor his assistants were familiar with the acts that regulate the use of pesticides. The only information available in the agricultural office during our visits in 1984 and 1986 was a Book of Law printed in 1962. This included under Pharmacy and Poison Act (Cap. 244) a list of toxic chemicals. However, the list did not include more than a fraction of the multitude of chemicals, particularly pesticides, that are in use today.

In 1981 work was initiated on a "Pest Control Products Act" meant to regulate the production, storage, packing and sale of chemical pesticides (Christiansson et al., 1984). It was going to replace the nine separate acts that then regulated the handling of chemical products

However, in April 1986, comprehensive legislation concerning the handling and storage of pesticides and other toxic substances in Kenya was still lacking.

A step to improve the situation was taken by the government when it recently issued a directive that all agro-chemicals must be labeled with instructions in both English and Kiswahili.

9. SUMMARY, DISCUSSION AND CONCLUSIONS

During the last few decades the market for agricultural chemicals has increased phenomenally in the Third World. At the same time an increase in health problems related to the use of chemical products is reported. These facts are well documented, and a few of the many sources were cited earlier in the text. The general trends are reflected in the results of the studies in Endarasha and Othaya.

Now, are these locations representative of rural Kenya? Yes, we have found that they are at least representative of those areas with a high population density and intense smallholder agriculture with coffee, vegetables or root crops as the principal cash crops. This was shown by the main study which dealt primarily with patterns of land use and socio-economic conditions (Anon., 1977-1988).

When the main study was planned in 1977, the pesticide aspects were not included. Those questions were successively brought into the picture and given particular attention during the period 1982-1986.

Sale and use of pesticides in Endarasha and Othaya occured at that time without any control by the government. The availability of products and the fixation of prices seem to have rested mainly with the producers and the dealers. The producers, frequently transnational chemical companies, considered and still consider the smallholder sector in Kenya as an interesting market. Thus, there is a large variety of pesticides available in shops in the rural areas. In their marketing the transnational companies and their local sales offices stress the use of chemicals as the most important requirement for successful and modern farming. This picture is also communicated to the extension staff of the Ministry of Agriculture.

The extension staff, many of whom were trained long before pesticides played the role they play today, get part of their information about pesticides via the ministry's common channels, the Training and Visit system (T&V), but it seems as though they get the major part of their information via the companies that sell the products concerned. The extension workers themselves have limited possibilities of judging the properties of the different products. Mostly, their role is to transmit the information printed in marketing folders or on the packages. It is impossible for them to keep themselves informed about the very large variety of pesticides that is available on the Kenyan market. New products turn up constantly. The recommendations they can give the farmers appear arbitrary and the advice given by different extension workers differs in many respects. It may be that the recommendations primarily reflect which company is the most successful in its marketing just at that time.

What chance do the farmers have to make a choice among the products, to judge which are the most suitable, to know how they shall be handled, etc? In these respects the farmers have absolutely no formal training. Some are illiterate. There is no way for them to keep informed about the variety of products and their properties. As has been shown in this study they get information to some extent from the agricultural extension staff, via the T & V system, but primarily through the pesticide dealers. Decisive for their choice of products are the cost aspects, but, possibly, just as important is whether or not the product they want is available at all when they need it. This is one of the explanations for the observation made that the farmers frequently were not using the products recommended by the extension staff.

In this situation it is easy to understand the impact of advertisments and posters found everywhere in the rural areas. During our studies we noted numerous varieties of this intensive marketing that we fear leads to overuse, the use of the wrong products and improper handling of the pesticides. Improper handling is frequently due to non-existent or incomplete information on sacks and containers. It is not uncommon for pesticides to be repacked and sold in small quantities, particularly when sold in open-air markets or small private shops. During this procedure the warning texts disappear. We saw several examples of this during our study. Tins and bags that have contained pesticides are being reused for other purposes in households. This is not because the farmers have not been warned about this by health and agricultural officers but because they can not afford to throw away usable vessels and containers.

What has been said above in combination with the insignificant use of protective equipment - due to the costs - is part of the explanation for the health problems that can be related to the pesticide use shown in Figure 5. Unfortunately, a common attitude among dealers and health and agricultural officials is that health disorders for the above reasons are not a major problem. This is, however, contradicted by some recent statistics reporting hundreds of deaths in Kenya related to the use of agricultural chemicals (Orlale, 1989).

There is no doubt that in the areas included in our study, coffee, responsible for 50 per cent of Kenyas export income, is the crop most heavily treated with pest-icides. Reasons: Coffee production within the smallholder sector in Kenya increased during the 1980's. Coffee is sensitive to attacks and - possibly the most important factor - the quality control when farmers deliver their coffee to the coffee-factories is strict. (The coffee is sprayed against fungus diseases between five and eleven times a year. The fields are also treated two or three times a year against weeds and the crop is treated two or more times every year if insect attacks occur. Even in this respect the household economy plays a decisive role in terms of how often the farmer can afford to spray his crop).

Most farmers can not afford to spend large amounts of money on pesticide application on the subsistence crops. However, pesticides are used on these crops, particlarly if a surplus is produced and stored for later sale on the local market. Our study shows that the use of pesticides on smallholdings in Endarasha that produce virtually no export crops was less frequent than in Othaya where coffee is the most important cash crop. However, all farmers in Othaya and most in Endarasha use pesticides.

All the farmers using pesticides clearly have a positive attitude to them. Although many farmers get health problems in connection with spraying, the knowledge of effects on humans and on the environment is very limited both among the farmers themselves, the agricultural extension workers and health workers. Many farmers even seem to accept the health problems that follow from pesticide use. "That is the cost of improved harvests." But what are the social and economic implications of many members of rural households being temporarily or permanently debilitated or made incapable of work?

Are there any alternatives to chemical pesticides in modern smallholder agriculture in Kenya? There is very little debate on this among agricultural and health authorities. Options to reduce chemical pesticide dependency certainly exist. Among those are increased use of crop rotation techniques, development of resistant crop varieties and the use of natural predators.

One of the approaches that excite researchers today is the use of products from the "neem" tree as crop pesticides. The neem tree grows in Africa and South Asia where it has been used in pest control for hundreds of years. Another option is "lilupala", a root which is cultivated in Tanzania and used by peasants as a substitute for DDT. According to the users it is just as efficient as commercial pesticides (Friis-Hansen, 1987).

Pyrethrin, extracted from chrysanthemums, is the basis of an important family of pesticides - the pyrethroids. They are already widely used and less toxic to humans than many of the chemical products.

To increase agricultural output and at the same time to care for the health of the farming population and the environment require massive research on the benefits and problems of pesticide use. Alternative ways have to be sought to avoid hazards and to develop biological pest control methods. Strict government-supervised programmes must be established to control insects and crop pests but programmes must also be established to monitor the effects (including health effects) of the chemical pest control. Tough legislation is necessary coupled with rigorous enforcement. It is also imperative that the government ensures that health and agricultural officers as well as farmers are well versed in the proper use and handling of pesticides and aware of alternative crop protection methods.

ACKNOWLEDGEMENTS

It was only with the assistance of a very large number of persons that this study could be carried out. We particularly want to direct our thanks to the District Officers in Othaya and Kieni West Divisions who gave us permission to carry out the investigations and who supported our studies in many ways. We also appreciate the invaluable assistance provided by Mr Peter Mwangi, assistant Chief, Thuti and Mr Paul Njoroge, Chief, Endarasha. Thanks are also due to the divisional staff of the Kenya Ministry of Agriculture and the Ministry of Health in Othaya and Kieni West as well as to the staff of the Othaya Hospital. Further thanks are due to all those farmers and dealers in chemicals who patiently answered our questions and to the numerous Swedish field course participants and their Kenyan field assistants who made a survey of this kind possible. The author is particlarly grateful to Dr. Lori Ann Thrupp, University of California, who read the draft manuscript and suggested many valuable revisions. The assistance by Professor Nils Kautsky, University of Stockholm, Professor Wilhelm Umaerus, Swedish University of Agricultural Sciences, and Dr. Bo Göhl, SAREC, who read and commented on draft versions of the manuscript, is also gratefully acknowledged.

REFERENCES

Acland, J.D. 1971. East African Crops. FAO/Longmans, London.

Ahmad, T. 1987. Waging War on Rice Pests. The Panos Institute, London.

Anonymous, 1977-1988. Field reports 1977-1988. Departments of Physical and Human Geography, University of Stockholm, Stockholm.

Berglund S. and Johansson, D. 1982. Miljöpåverkan i odlingslandskapet. (Environmental impact in the agrarian landscape.) LT:s förlag, Stockholm.

Boza Barducci, T. 1972. Ecological Consequences of Pesticides used for the Control of Cotton Insects in Canete Valley, Peru. In: Farvar, M.T. and Milton, J.P. (eds.). The Careless Technology. Natural History Press, New York.

Bull, D. 1982. A Growing Problem. Pesticides and the Third World Poor. Oxfam, Oxford.

Bull, D. and Myers, D. 1984. Reglera gifthandeln (Establish control of the pesticide trade). Rapport från SIDA, Nr 5 1984. SIDA, Stockholm.

Christiansson, C. and Wåhlin, L. 1982. Geographical Studies in Highland Kenya 1982. Departments of Physical and Human Geography, University of Stockholm, Stockholm.

Christiansson, C. and Strand, B. 1983. Geographical Studies in Highland Kenya 1983. Departments of Physical and Human Geography, University of Stockholm, Stockholm.

Christiansson, C., Ekberg, N., Strand, B. and Wåhlin, L. 1984. Geographical Studies in Highland Kenya 1984. Departments of Physical and Human Geography, University of Stockholm, Stockholm.

Christiansson, C., Kinlund, P., Larsson, M., Wåhlin, L. and Östman, P. 1986. Geographical Studies in Highland Kenya 1986. Departments of Physical and Human Geography, University of Stockholm, Stockholm.

Cox, P. 1985. Pesticide Use in Tanzania. ODI and Economic Research Bureau, University of Dar es Salaam, London and Dar es Salaam.

Ekberg, N., Kinlund, P., Larsson, M. and Wåhlin, L. 1985. Geographical Studies in Highland Kenya 1985. Departments of Physical and Human Geography, University of Stockholm, Stockholm.

Forget, G. 1989: Pesticides: Necessary but dangerous poisons. *IDRC Reports* 18:4-5.

Friis-Hansen, E. 1987. Changes in Land Tenure and Land Use since Villagization and their Impact on Peasant Agricultural Production in Tanzania. CDR Research Report No 11, IRA Research Paper No 16, Copenhagen.

v. Hofsten, B. and Ekström, G. 1985: Control of Pesticide Applications and Residues in Food. Swedish Science Press. Uppsala.

McCracken, J.A.and Conway, G.R. 1987. Pesticide Hazards in the Third World: New Evidence from the Philippines. Gate Keeper Series No SA2. IIED, London.

McCracken, J.A., Pretty, J.N. and Conway, G.R, 1988. An Introduction to Rapid Rural Appraisal for Agricultural Development. Sustainable Agriculture Programme. IIED, London.

Ngugi, D.N., Karau, P.K. and Nguyo, W. 1978. East African Agriculture. Macmillan, London.

Orlale, O. 1989. Pesticides at home and work. *IDRC Reports* 18:10-11.

Pimentel, D. and Edwards, C. A. 1982. Pesticides and Ecosystems. *BioScience* 32:595 -600.

Singh, J.P. 1983. Crop Protection in the Tropics. Vikas Publishing House, New Delhi.

Weir, D. and Schapiro, M. 1983. Circle of Poison. Jord-Eco Förlag, Luleå.

Woller, R. 1983. Endurance Test in the Tropics. Hoechst News 79, Frankfurt.

Linking the Natural Environment and the Economy; Essays from the Eco-Eco Group,
Carl Folke and Tomas Kåberger (editors)
Second Edition.
1992. Kluwer Academic Publishers

CHAPTER 12

Multinational Firms and Pollution in Developing Countries[1]

by

Thomas Andersson

Industrial Institute for Economic and Social Research (IUI)
Box 5501, S-114 85 Stockholm, Sweden

Competition between developing countries that hope to host multinational enterprises should stimulate an efficient pattern of pollution intensive direct investment combined with an optimal level of pollution abatement. The reasons for environmental neglect in developing countries are likely to be found in imperfections in the international capital markets, lack of information and "government failure". International cooperation may be needed to put pressure on individual governments not to downgrade environmental quality below its social value.

1. INTRODUCTION

Many economic activities affect the environment adversly. The impacts tend to be external in nature, which means that there are no monetary transactions compensating for the gains or losses in welfare. There is, in other words, no market mechanism in operation. Economic decision-makers take environmental impacts into consideration only if regulated by law or action by those affected.

This century has witnessed considerable environmental degradation in many areas of the world. To some extent the damage is irreversible, meaning that it will not be possible to undo at a reasonable cost. Meanwhile, in the industrialized countries there has been an unprecedented expansion of industrial output, and a much enhanced standard of living as measured by conventional welfare indices. Today, there is a growing appreciation of environmental quality and steps are being taken to secure sound management of the environment. In the developing countries by contrast, little attention is paid to the matter, and environmental degradation is increasing in many places.

A society should, ideally, take environmental impacts into account in accordance with their social values. We here refer to the assignment of such values as *cost-benefit analysis,* in which the effects are *identified, quantified* and *valued* in social terms. In practice, this task encounters substantial difficulties. For example, environmental impacts tend to permeate ecological systems, where the final outcome is uncertain and hinges on a synergy of factors. Such difficulties are sometimes used as an argument against the use of cost-benefit analysis. This is, however, the only method currently available to relate environmental values to others with which they inevitably compete as well as co-exist.

Given the complexity of environmental impacts, properly managing the three steps of a cost-benefit analysis often requires cooperation between experts in different fields, e.g. physicists, biologists, physicians, toxologists, sociologist, and economists. This is particularly the case when the sum of many marginal impacts develops into a synergetic degradation which surmounts that of the individual constituents. Given the speed of current environmental degradation, developing an adequate basis for environmental management is an urgent task for researchers across a wide scientific spectrum.

For the implementation of sound environmental management, the understanding of social values must be transformed into political will, so that governments do not give in to conflicting market interests. Comprehensive information is vital to the willingness of politicians as well as individuals to stand up for environmental values. Insufficient or too costly information will be accompanied by neglect of protection that is desirable, as well as misguided attempts to institute protection that is not needed.

It is well known that multinational enterprises (MNEs), which operate across the boundaries of nation states, may relocate pollution intensive activities, as well as play governments in different countries against each other. Against this background, the present article examines factors expected to *influence the level of pollution abatement required of multinational enterprises in developing countries*. I add to most previous studies of MNE-host country interaction by considering that *developing countries are each potential locations for many investment projects, and can be expected to compete with each other for the gains.*

After an introductory discussion of direct investment and its role in pollution-intensive activities, the article uses a *game theory framework to argue that interaction between MNEs and competing host countries does not result in too much environmental degradation.* On the contrary, such interaction serves to establish an efficient pattern of investment and a socially optimal level of pollution abatement. The fact, however, that excessive pollution can be observed in many developing countries takes us back to the importance of *information* and *the behaviour of governments*. The final section summarizes and discusses the policy implications.

One troublesome aspect of environmental impacts, which is not much discussed in

this study, concerns their transboundary character. It is well-k
global irrationality in the uncoordinated behaviour of individual
common resources, such as the oceans or the atmosphere (
Dasgupta, 1976). Likewise, spillovers may be transmitted through e.g.
the atmosphere. Here, we are primarily concerned with effects that are co[illegible] the
country in which investment is undertaken.

2. DIRECT INVESTMENT AND POLLUTION

A MNE is a firm that owns and controls subsidiaries in other countries. The ownership and control is realized through direct investment, which consists of a majority equity share.[2] As is common today, our point of departure is the eclectic or "OLI"-framework (Dunning, 1977). According to this, direct investment is motivated by ownership advantages, combined with advantages of internalizing them within a firm and locational factors that make a specific country the optimal location.

For a MNE, as for any firm, there is an incentive to internalize revenues but keep costs external. Pollution abatement merely represents one example of costs. A relatively lax environmental protection framework constitues a cost incentive for any firm to locate pollution intensive activities in that country. It may therefore be asked whether there is any reason to pay special attention to MNEs in regard to pollution in developing countries. There are at least three arguments brought forward as to why this might be;

- Firstly, the broad scope of these firms' activities in pollution-intensive industries results in an impact on the environment.
- Secondly, their dominance in technology influences industrial processes, including environmental impacts.
- Thirdly, MNEs function across the boundaries of nation states and are in a position to bargain with individual countries, on issues such as pollution, to an extent which domestic firms are not.

The rest of this section discusses the first two reasons. We then turn to the third possibility, which is the issue of focus in this article.

Two kinds of pollution can be distinguished: production pollution and consumption pollution.[3] The former encompasses material-source (due to extraction and transportation of natural resources) and process pollution. The latter includes product and residual pollution (related to the disposal of products). Allthough we are directly concerned only with production pollution, in practice the two are often intertwined.

There are several ways to identify pollution-intensive industries. Table 1 reports those that have been the hardest hit by pollution compliance costs in U.S. manufacturing. As can be seen, four industries pay some three-fourths of all capital expenditures for pollution control. Using a range of criteria, UNCTC defines roughly the following industries as pollution intensive; chemicals, agro-industry, aluminium, iron and steel,

Table 1. New plant and equipment expeditures for pollution abatement, 1973-84 (in billions of dollars).

	1973	1974	1975	1976	1977	1978	1979	1980	1981	1982	1983	1984
Mineral processing	0.70	0.78	1.05	0.97	0.90	0.79	0.94	1.01	0.83	0.80	0.56	0.74
Chemicals	0.42	0.54	0.76	0.88	0.82	0.62	0.62	0.69	0.79	0.68	0.61	0.58
Pulp and paper	0.31	0.38	0.47	0.43	0.37	0.33	0.43	0.46	0.45	0.35	0.37	0.54
Petroleum	0.54	0.75	1.21	1.10	1.04	1.13	1.26	1.54	1.68	1.57	1.46	1.28
Total high-pollution sectors	1.97	2.45	3.49	3.38	3.13	2.87	3.25	3.70	3.75	3.40	3.00	3.14
All manufacturing	3.17	3.62	4.66	4.49	4.32	4.30	4.77	5.35	5.36	4.78	4.26	4.53
High-pollution-sector expenditures as percentage of all manufacturing	62.1	67.7	74.9	75.3	72.5	66.7	68.1	69.2	70.0	71.1	70.4	69.3

Source: Environment Economics Division, 1986

motor vehicles, nonferrous metals, petroleum and coal products, pulp and paper, and stone, clay and glass products.[4]

How important are MNEs in pollution-intensive industries? Broadly speaking, the developing countries expanded their share of the world's industrial value added in such industries between 1973 and 1980. The share of direct investment varied greatly, and industrial expansion was in several cases achieved primarily by domestic industry. At the same time it has been pointed out, e.g. by Dunning & Pearce (1981), that there is considerable direct investment in developing countries in practically all pollution intensive industries. This is particulary the case for chemicals, including petrochemicals, pharmaceuticals, paints, plastic products, fertilizers and pesticides. Mining, petroleum extraction and refining, agri-business, refining of heavy metals, woods and paper processing and motor vehicles are also major destinations for direct investment.

The consequences of direct investment in pollution-intensive industries must be evaluated with respect to whether direct investment substitutes for domestic production or adds to it. Here, we encounter the second argument as to why MNEs deserve special attention. The sheer size of direct investment does not alone allow us to determine its importance. The crucial question concerns how the MNEs' technology compares to that of domestic firms, and how the latter is influenced by the former.

MNEs play a primary role in the dissemination of industrial technology in general, and environmental management technology in particular. The transfers may be embodied in new investment, training, trade letters, licensing of control processes, direct sales of products or services, etc. The availability of advanced environmental management must, of course, be separated from its application. Advanced pollution control is generally not desired by affiliates, and is seldom forced upon them by parent companies. MNEs tend to pollute more in developing countries than at home, but less than domestic firms.[5]

Given the scope of MNEs in pollution-intensive industries and their importance in technology transfers, their role in environmental degradation is a matter of great concern. However, as they typically use technologies that pollute less than domestic firms, and given that pollution-intensive industries sooner or later will flourish in developing countries (to the extent that they constitute the optimal locations), it is inappropriate to focus on the role of direct investment. The object of study should rather be pollution-intensive activities in general. Thus, we are left with the third argument for why MNEs require special attention - their ability to play governments against each other.

3. BARGAINING WITH ENVIRONMENTAL PROTECTION

How do competing host countries behave vis-à-vis MNEs that have pollution-intensive investment projects? This question is here analyzed using a bargaining framework - an approach which goes back to Vernon (1971) and Moran (1974). Both MNEs and host countries seek to gain as much as possible from direct investment at each point in time. A host country offers a MNE a certain level of "taxation" and "environmental protection". The former essentially concerns payments of foreign exchange, the latter pollution abatement.[6]

Let us assume that firms act as rational maximizers of profits and host countries of socio-economic gains, that the latter can discriminate their policies vis-à-vis MNEs, and that the two parts interact in a non-cooperative game with complete information over an infinite time horizon. This implies, among other things, that environmental impacts can be identified, quantified and valued in monetary terms. A MNE possesses specific advantages which enable it to generate a profit from direct investment in a developing country. The profit can, from a host country's perspective, be transformed into foreign exchange. In addition to the profit, the activity results in negative impacts on the environment. The profit and the cost of the impact are somehow to be shared between the MNE and a host country.

Now, in contrast to most previous studies, consider that alternative host countries compete for gains from direct investment. Given that the net gain from a certain project is slightly greater in one country than in others, it is possible to demostrate the existence of a unique (subgame perfect) equilibrium (Andersson, 1991a). The net gain of a host country is determined by the superiority of its investment opportunities relative to those of the second best alternative, and by the "mobility" of investment. A host country must keep its corporate income tax sufficiently low to compensate a MNE for what it would earn from direct investment in the second best country instead. In order to gain something, that country is prepared to forgo all its gains to attract the project. The sunk cost and the discount factor, which are associated with the mobility of investment, determine what would be lost by relocating production.

Thus, the country with better investment opportunities than its competitors gives up only part of its gains, and still attracts investment. The smaller the difference between it and the second best alternative, the less gain it is able to capture. Countries which represent the second or third best opportunities stretch their offers until they exhaust all their potential gains, but still are unable to obtain a project. *How much* a host country earns consequently depends on its investment opportunities relative other countries, and the mobility of investment. *What* it gives up depends on its evaluation of the need for foreign exchange versus environmental quality. It must balance one more dollar in pollution abatement against a dollar less of foreign exchange. Meanwhile, the benefit of pollution abatement is given by the social value of environmental protection.

Nothing in this situation motivates any deviation between the marginal cost of environmental protection and the marginal benefit of cleaning. *Effective host country competition does not lead to socio-economically inefficient environmental degradation.* With advantages in technology and management, and an experience of operations in industrialized countries, MNEs should have a lower marginal cost curve for pollution abatement than the domestic firms in a developing country. Given that a host country requires a lower marginal cost for pollution abatement than the home country, a MNE is more polluting here than it is at home, but less so than domestic firms.

It can be noted that countries of origin, like host countries, do not require pollution abatement across the board, but also bargain with individual firms. One may therefore ask whether the level of pollution abatement at home is restricted by a firm's possibility to move abroad. In analogy with the above analysis, the answer is no. The home country earnings are restricted by a MNE's possibility to move abroad, just as the earnings of one host country are restricted by the level that could be achieved in another country. The earning are determined as a trade-off between taxation and environmental protection. The cost of cleaning should not deviate from the social benefits at the margin.

In sum, the net gain enjoyed by a host country is dependent on the quality of its investment opportunities relative to those of the second best alternative, and by the mobility of investment projects. The level of environmental protection is established through a trade-off with foreign exchange earnings, wich means that the cost should equal the benefit of foregoing further degradation at the margin. What are the implications for the pattern of direct investment?

4. THE PATTERN OF DIRECT INVESTMENT

In recent years, new laws, regulations and private efforts have enforced stricter environmental protection, and raised the cost for pollution-intensive industries in industrialized countries. This might be expected to give rise to a transfer of such activities to developing countries where regulation remains lax or even non-existent. This idea was first brought forward by Walter (1972), who predicted environmental pressures would promote a gradual shift of pollution-intensive activities towards countries with lower control costs, meaning a shift from high-income to low-income countries. MNEs would play a primary role in this process. Pearson (1976) predicted that the greater the differences in environmental control costs, the less tied industries are to specific inputs or markets, and the more successful past foreign operations have been, the greater the amount of relocation.

A number of empirical studies have examined the extent to which these prophesies have been realized. Walter (1975) himself found some evidence of geographical mobility when projects had been blocked at home, but not much evidence at the aggregate level that the pattern of direct investment would have been seriously affected by environmental considerations. Later studies have not found much relocation motivated by environmental control costs, although there are questions concerning some chemical industries.

The effects of relocation to developing countries in the latter case mainly concern two industries, those producing highly toxic products such as asbestos, benzidine dyes and pesticides, and those producing heavy metals such as copper, zinc and lead. In both cases, new investment has tended to be located in developing countries when plants are closed in home countries. Leonard (1988) points out that those susceptible to "industrial flight" are ailing industries that have experienced slow growth in domestic demand, and which are "not likely to contribute in any significant way to the development of countries trying to build their industrial base".

On the whole, it can be concluded that *there is fairly little evidence of environmentally motivated relocation of pollution-intensive direct investment.* One response has been that "environmental costs do not matter". Gladwin and Welles (1976) expressed scepticism concerning the potential for relocation, arguing that the elasticity of investment is low with respect to environmental control costs. Leonard (1984) argued that pollution intensive industries have adapted through technological innovations rather than relocation across national boundaries. Pearson and Pryor (1978), on the other hand, maintained that the issue remains controversial until definite estimates are available.

As discussed by UNCTC (1985), there is a close correlation between the stringency of environmental policy and income level of a country. Meanwhile, most direct investment is obtained by the relatively high income "newly industrialized countries", in which environmental protection has been growing the most.[7] While this confuses a simple comparison between the location of investment and the level of protection, it is compatible with our notion that countries bargain on the basis of their investment opportunities. Differences in host country earnings may be the natural outcome of competition between dissimilar countries, while inter-country differences in environmental protection reflect differences in the social value of pollution.

To the extent that our model framework is applicable, there will be no relocation of direct investment from countries with superior investment opportunities to those with less favourable ones. Environmental impacts influence the location of projects only if they change in what country a project's net value is the largest. In that case, another country can afford to offer a MNE the largest net profit. Such a change must stem from differences either in assimilative capacity, or in the valuation of given effects.

It is sometimes argued that developing countries are natural dumps for pollution. Under this argument, ecological balance is a luxury not affordable at the present level of development. Faced with serious and acute problems requiring urgent investment, e.g. hunger, illiteracy, unemployment, rural-urban imbalances, and chronic poverty, these countries can not afford much environmental protection. The socially optimal level of protection would be lower than in the industrialized countries.

It cannot be taken for granted, however, that developing countries are optimal locations for pollution-intensive activities. There is, first of all, no evidence of a relatively high assimilative capacity for pollutants. While a hot, dry climate may reduce the impact of some effluents, and high rainfall may reduce the impact on air quality, temperate industrial countries have a higher assimilative capacity for many other effluents than tropical developing countries. Biological and chemical compounds generally spread more easily in the latter, people are less resistant to diseases, etc. Secondly, pollution may reduce the usefulness of productive resources, and developing countries have less capacity to counteract such effects. For example, the stripping of forests may destroy land for all future use, or phosphate mining may make entire islands uninhabitable. Thirdly, industries tend to be concentrated in developing countries, and further expansion is unlikely to be more dispersed. While some decades ago industrial centres in developing countries tended to have a lower pollution intensity than thouse in industrialized countries, the situation is rather the opposite today.

Thus, the assimilative capacity of developing countries makes an indiscriminate transfer of pollutants to them unwise, even if it may favor a transfer in some cases. The differences in the valuation of environmental quality may motivate some additional ones, but not when the outcome is a disminished productive capacity for the host country. The limited evidence that is currently available regarding relocation of direct investment may indicate that the net of the differences in assimilative capacity and valuation of effects does not generally alter the ranking of economies in terms of investment opportunities.

In sum, the country with the best investment opportunities offers the most favourable conditions, and therefore obtains a project. Environmental effects influence the pattern of direct investment only to the extent that their impact on the net value of a project is sufficient to alter the ranking of potential host countries. Of course, these results hinge on the assumption that there are no imperfections in the model. In practice, information may be incomplete and agents do not always behave in an economically rational manner. The ensuing section discusses factors that are believed important for explaining deviations from the model in the real world.

5. FACTORS AFFECTING THE MODEL RESULTS

We have argued above that the country which yields the highest net value of investment is able to secure the project and that the marginal cost of cleaning equals the marginal social benefit. Given a limited assimilative capacity in developing countries, our framework is consistent with relatively little relocation of pollution intensive direct investment. On the other hand, this might also be due to a lack of stringent regulations in industrialized countries. As control costs increase, more relocation of pollution-intensive direct investment to developing countries will occur.

Moreover, there are serious deviations from the model in the real world. In the face of considerable environmental problems, many developing countries have virtually no pollution control at all. Certainly, the social benefit of pollution abatement is not zero. The air quality in Mexico City, Beijing or Taipei, or the irreversible destruction of the rain forests in Southeast Asia, Central Africa and Latin America, hardly bear witness to socially optimal environmental protection. Deviations could appear in *i)* the balance the marginal cost of protection and the scarcity cost of foreign exchange, *ii)* the balance between the marginal cost and the social benefit of cleaning, or *iii)* the motives of governments. In this section, we discuss these three categories.

i) The developing countries are relatively scarce in capital and abundant in labour, so that the marginal rate of return to capital should be relatively high. Since savings require less consumption, and consumption is low already, there are limits to what can be saved domestically. Investment should instead come from the industrialized countries. However, the risk of debt repudiation, limited taxing power of the host country government or imperfections related to the supply of loans (such as the risk of panic among creditors) often lead to severe credit rationing. Likewise, host country policies that interfere with the ownership of affiliates ("nationalization") may prevent the undertaking of direct investment (Andersson, 1991a).

With impediments to portfolio as well as direct investment, capital remains scarce in many developing countries. Other contributing factors are: barriers to developing country imports in industrialized countries and their possible escalation in the future; debt burdens accumulated when interest rates were low and that become worse with rising interest rates; budget deficits and soaring inflation resulting in overvalued exchange rates. Consequently, many developing countries today find themselves severely constrained with regard to foreign exchange, and the marginal rate of return to capital remains higher than in the rest of the world. Other resources, such as environmental quality, are downgraded. Viewed from a global perspective, this represents an inefficient allocation of resources. This is particularly evident when benefits of pollution abatement accrue to other countries.

ii) It has been assumed that the cost and social benefit of pollution abatement are equal at the margin. In practice, there may be imperfections in information concerning both the cost and the social benefits of abatement. In particular, it is often difficult to identify, quantify or put a value on environmental impacts. The increased protection in industrialized countries has been paralleled by a great improvement in information. The developing countries have, by contrast, limited technical, economic and administrative expertise to trace the effects, or check the compliance with pollution control. Some have passed fairly rigorous laws, but they are seldom enforced.[8] Due to their relative lack of resources for information gathering, the developing countries suffer greater environmental risks than the industrialized.

An individual's perspective on risk may differ from that of the society as a whole. Under fairly strict assumptions, Arrow and Lind (1970) argued that, if risks are borne by government, risk-dispersion implies that society should be risk-neutral. In reality, imperfect risk-dispersion may violate this proposition. The willingness of risk-averse individuals to pay extra in order to retain certain options for the future can theoretically be represented by an "option value" (Weisbroad, 1964). Option values may be either positive or negative, however, and may affect the valuation of impacts in either direction.

More straightforward implications follow from a lack of markets that relate future goods and services to current values. Ecological systems represent imperfectly known resources which tend to take the form of collective goods or factors of production, meaning that there are no private incentives for their preservation.[9] While the outcome often is uncertain at the outset, the damage may be irreversible. For the irreversible destruction of currently unknown values one can assess a "quasi-option value". However, it is impossible to assign it a precise estimate (Fisher, 1981).

Concerning the valuation of environmental impacts there are, first, difficulties in accurately estimating recreation values, option values and bequest values, among others. While there is no automatic market mechanism for their articulation, one technique is to use implicit markets in which such values are linked to the consumption of ordinary goods. An alternative is to create an artificial market, which normally means asking consumers about their willingness to pay. The difficulties that pertain to both techniques are particularly severe in developing countries, where markets often are distorted and malfunction in various ways.

There is not only uncertainty concerning present consumer preferences. The income elasticity of environmental preferences tends to be larger than one, so that environmental quality can be expected to be valued more highly in the future, but the size of the increase is unkown today. Moreover, environmental preferences tend to be unstable and may be strongly influenced by the kind of information that is available.

iii) A government that maximizes social welfare over a sufficiently long time horizon could account for socio-economically optimal investment in the provision of information on enviromental cost and benefits. Remaining risks should, ideally, be reflected in option values and quasi-option values. Adjustments could be made to take into account an expected upgrading of consumer preferences in the future. In practice, many governments in the Third World have often proved to be no more inclined than private agents to spend resources on investigating or preventing environmental degradation.

This brings us to the matter of "government failure". With the school of public choice, it has become widely questioned whether economics and politics can be separated. A government need not maximize social welfare, but may act according to a self-interest that differs from that of society as a whole. A society is heterogeneous and regimes (whether autocratic or democratic) may choose to base power in certain, influential groups rather than less articulate ones.

Like other social cost and benefits, environmental degradation affects households or individuals unevenly. Accepting pollution consequently involves a decision of which groups are to have their welfare cut. This can be made less politically painful by targeting the effects towards groups that are the least likely to be aware of, and/or protest, their exposure to risk. One aspect of this problem is the prevention of free publication and criticism of environmental mismanagement. Particularly societies without free elections tend to prevent competing movements from capitalizing on the need for changes in public intervention to prevent environmental degradation (Bojö et al., 1990). Instead, politicians seeking to maximize their self-interest readily make use of foreign exchange earnings.

Imperfections in the capital markets and in information, and "government failure" may upgrade the demand for foreign exchange relative to environmental protection. Host countries then choose to capture too much of their gains from direct investment as foreign exchange earnings, and allow too much pollution. Of course, a country that downgrades environmental quality more than other, competing countries, may attract direct investment although its net value is not at a maximum. In this case, the investment pattern runs contrary to an efficient inter-country allocation of resources. As made clear, such an outcome does not derive from host country competition per se, and it is not inherent to the behaviour of developing countries vis-à-vis MNEs. The causes are to be found in factors such as those discussed in this section, and the resulting bias against pollution control applies more or less to all economic activities, whether domestic or foreign-owned.

6. CONCLUDING REMARKS

Multinational enterprises (MNEs) play an important role in pollution-intensive activities. The present analysis has considered host country competition for investment projects which give rise both to profits and negative external effects on the environment. Offering lax environmental protection requirements (accepting pollution) is a policy tool to attract such pollution-intensive direct investment.

Given complete information, the net gain of a host country depends on the quality of its investment opportunities in relation to those of the second best country, and the mobility of investment. The marginal cost of enviromental protection equals the marginal social benefit. Possible differences in the level of protection between countries do not influence the pattern of direct investment, unless they reflect differences in the capacity to assimilate pollution or in the valuation of given impacts. Whether developing countries do represent an optimal location for pollution-intensive activities is an open question. The limited relocation observed in practice may indicate that in most cases they are not.

While the above results are consistent with many observations in the real world, they are at odds with the common absence of effective protection in the face of rapid environmental degradation. It has been suggested that the causes are to be found not in host country competition, but in imperfections in the international capital markets, in access to information and "government failure".

These findings may make it seem ill-advised to opt for international cooperation regarding environmental policies as is currently done by OECD and the United Nations, for example. That conclusion is premature. First, we have not dealt with the common resources. When the benefits of production accrue to one country, but the costs are carried by all countries together, optimal protection may require cooperation. As investigated in Andersson (1991b), neglect of environmental protection by one country may then reduce the level of pollution abatement which is desirable in other countries. Secondly, stimulation of technological advancement in pollution abatement, and its dissemination among countries, may motivate forced internalization of environmental impacts within firms throughout the globe. However, inter-country cooperation is not motivated by host country competition per se.

Rather than striving for universal standards, international cooperation should aim to put pressure on individual governments not to downgrade environmental quality below its social value, and it should be designed in the light of this goal. To achieve this end it has to be part of a broader strategy. If environmental values are not to be the prey of irresponsible governments, polluters have to be confronted with the social costs they inflict whether they are of domestic or foreign origin.

NOTES

1. The article is a revised version of Chapter 5 in Andersson (1991a).
2. Direct investment is identified by the International Monetary Fund as "investment acquiring a lasting interest in a foreign economy, the purpose being an effective voice in the management of an enterprise".
3. Walter (1975) discussed the following sources of pollution; gaseous discharges, liquid and solid discharges, thermal discharges, noise, radiation, disposal of solid waste, degradation of natural scenery and terrain, including the elimination of recreational opportunities, endangering of wildlife species and congestion.
4. It has been estimated that the sum of direct and indirect costs of environmental controls average 39 to 52 per cent of the total costs in some of these industries. Adding the consumer industries, we have tobacco, motor vehicles and industrial and farm equipment industries. See UNCTC (1985, p. 35) for background data and further references.
5. See UNCTC (1985) for references. Of course, there are differences between individual MNEs. Gladwin (1977) investigates firm-specific factors associated with environmental concern.
6. Much environmental regulation does, in fact, come about via explicit or implicit negotiation between MNEs and national governments, not least in developing countries (cf. UNCTC, 1985). Whether environmental protection is designed as regulation, tax incentives, subsidies or a market for emission rights is not considered here. Our concern is the level chosen.
7. The "very poor" countries, which had the most lax environmental protection, accounted for some 70 per cent of the population of developing countries while receiving some 2.6 per cent of direct investments in 1980-82. This is only a slightly larger share than ten years earlier.
8. Brazil and the Republic of Korea switched towards more strict environmental and land-use regulations in the late 1970s. Nigeria, Kenya and Malaysia have also taken such steps.
9. Many plants and animals are extinct before they become known. Biologists can nerely estimate the number of species that is currently lost in tropical forests. According to some the figure amounts to 10 000 each year. As an indication of what potential benefits may be lost it has been noted that one quarter of the prescription drugs in the United States have been derived from plants (Farnsworth and Soejarto, 1985).
10. For a survey, see UNCTC (1985) pp. 75-84, where further references can be found.

REFERENCES

Andersson, T. 1991a. Multinational Investment in Developing Countries: A Study of Taxation and Nationalization. Routledge, London.

Andersson, T. 1991b. Government Failure: the Cause of Global Environmental Mismanagement. *Ecological Economics* 4:215-236.

Arrow, K.J. and Lind, R.-C. 1970. Uncertainty and the Evaluation of Public Investments Decisions. *American Economic Review* 60:364-378.

Bojö, J., Mäler, K.-G. and Unemo, L. 1990. Environment and Development: An Economic Approach. Kluwer Academic Publishers, Dordrecht.

Casson, M. 1982. The Economic Theory of The Multinational Enterprise: Selected Papers. Macmillan, London.

Caves, R. 1982. Multinational Enterprise and Economic Analysis. Cambridge University Press, Cambridge, U.K.

Dagsupta, B. 1976. Environment and Development. UNDP, Nairobi.

Dunning, J.H. 1977. Trade, Location of Economic Activity and the MNE: A Search for an Eclectic Approach. In: Ohlin, B., Hesselborn, P.-O. and Wijkman, P.M. (eds.). The International Allocation of Economic Activity: Proceedings of a Nobel Symposium Held in Stockholm. Macmillan, London. pp. 395-418.

Dunning, J.H. and Pearce, R.D. 1981. The World's Largest Industrial Enterprises. Gower, Farnborough.

Environmental Economics Division. 1986. Plant and Equipment Expenditures by Business for Pollution Abatement. *Survey of Current Business* 66:39-45.

Fisher, A.C. 1981. Resource and Environmental Economics. Cambridge University Press, Cambridge.

Gladwin, T.N. 1977. Environment, Planning and The Multinational Corporation. Jai Press, Greenwich.

Gladwin, T.N. and Welles, J.G. 1976. Environmental Policy and Multinational Corporate Strategy. In: Walter, I. (ed.). Studies in International Environmental Economics. John Wiley, New York.

James, J. 1981. Growth, Technology and the Environment in Less Developed Countries: A Survey. In: Streeten, P. and Jolly, R. (eds.). Recent Issues in World Development. Pergamon Press, New York.

Leonard, J.H. 1984. Are Environmental Regulations Driving United States Industry Over-seas? An Issue Report. The Conservation Foundation, Washington.

Leonard, J.H. 1988. Pollution and the Struggle for the World Product. Cambridge University Press, Cambridge.

Moran, T.H. 1974. Multinational Corporations and the Politics of Dependence: Copper in Chile. Princeton University Press, Princeton.

Pearson, C. 1976. Implications for the Trade and Investment of Developing Countries of United States Environmental Controls. United Nations Conference on Trade and Development, New York.

Pearson, C. 1982. Environment and International Economic Policy. In: Rubin, S.J. and Graham, T.R. (eds.). Environment and Trade. Allanheld, Osmun, Totowa. pp. 56-57.

Pearson, C. and Pryor, A. 1978. Environment North and South: An Economic Interpretation. John Wiley, New York.

Selten, R. 1975. Re-examination of the Perfectedness Concept for Equilibrium Points in Extensive Games. *International Journal of Game Theory* 4:25-55.

United Nations Centre on Transnational Corporations (UNCTC). 1985. Environmental Aspects of the Activities of Transnational Corporations: A Survey. New York.

United Nations Centre on Transnational Corporations (UNCTC). 1988. Transnational Corporations in World Development. New York.

Vernon, R. 1971. Sovereignty at Bay. Basics Books, New York.

Walter, I. 1972. Environmental Control and Patterns of International Trade and Investment: An Emerging Policy Issue. *Banca Nazionale Del Lavoro Quarterly Review* 100:82-106.

Walter, I. 1975. International Economics of Pollution. Macmillan, London.

Ward, B. and Dubos, R. 1972. Only one Earth. Penguin, London.

Weisbroad, B.A. 1964. Collective-consumption Services of Individual-consumption Goods. *Quarterly Journal of Economics* 78:471-477.

WCED, World Commission for Environmental and Development. 1987. Our Common Future. Oxford University Press, Oxford.

Linking the Natural Environment and the Economy; Essays from the Eco-Eco Group,
Carl Folke and Tomas Kåberger (editors)
Second Edition
1992. Kluwer Academic Publishers

CHAPTER 13

Environmental Conservation for Development in Central America

by

Johan Åshuvud

Conservation in Central America has recently taken the step from protection of wildlife to a wider concept of safeguarding the integrity of the environment in order to attain ecologically and economically sound development. The focus is on translating the concepts of natural resource management, conservation and sustainable development into practical programmes for action through local institutions. Conservation is not seen as a separate sector but as a concept unifying different sectors of society including government and non-governmental organizations. This article chronicles the implementation of the International Union for Conservation of Nature's (IUCN's) regional programme for Central America. The programme, which is based on the World Conservation Strategy, covers tropical rain forests, coral reefs, lagoon systems, coastal wetlands and mangrove forests, and initiates and coordinates projects that demonstrate the link between conservation and development. Environmental impact assessments, institution building, and human resources development are other components of the programme.

1. INTRODUCTION

Central America is blessed with an abundance of natural diversity. In approximately 540,000 square kilometers is found a remarkable range of cultures, ecosystems and wildlife. Unfortunately, the region is also experiencing economic crisis, civil strife, rapid population growth, and severe depletion of its natural resources and destruction of its ecosystems.

In Central America a new approach is developing, one that is aimed at the sustainable ecological economic development of the region. This approach is to a large extent influenced by the World Conservation Strategy (1980), developed by the

International Union for Conservation of Nature (IUCN). The article describes how this strategy is implemented in the region, and summarizes environmental projects initiated by IUCN through their Regional Office for Central America.

1.1 Conservation for Development; a New Approach

In the last few years the seven countries of Central America -Belize, Guatemala, El Salvador, Honduras, Nicaragua, Costa Rica and Panama - have begun to link their developement problems to the use of their natural resource base. There is a growing awareness that conservation can help resolve the long-term development problems of the region.

As a consequence, conservation in Central America now includes the wider objective of safeguarding the integrity of the environment, i.e. genetic richness, water balance, soil fertility, in order to attain sustainable development, and not simply protect species and wildlands from destruction.

The general objective of IUCN's programme in Central America is to demonstrate the link between management of natural resources and the sustainability of development investments and translate this into practical programmes for action implemented by local institutions. In this process, conservation is not seen as a separate sector of development, but rather as a concept unifying different sectors of society as well as governmental and non-governmental activities. The process takes much of its character from the World Conservation Strategy (1980).

The challenges facing natural resource planners are many. While the region has an impressive number of technically trained scientists, the mechanisms for translating knowledge into sound development decisions are poorly coordinated and unstable. Overcentralization and unwieldly bureaucracies are often to blame. The problem is aggravated by government structures that have a limited capacity to plan for the long-term survival and prosperity of their nations. The countries' development potential depends in part on the capacity of local institutions to successfully implement sustainable development projects.

1.2 Regional Partnerships

IUCN's role in the region has been defined jointly with governmental institutions and non-governmental organizations (NGO's) in each country. Decision-makers and resource users are reached at all levels of society. The international assistance of a technical and scientific nature provided by IUCN is often used by local and regional NGOs to obtain support for their work and by governments to negotiate for development aid. IUCN's major regional partner in Central America is the Tropical Agricultural Research and Training Center (CATIE) based in Turrialba, Costa Rica. This NGO includes all Spanish-speaking countries in Central America and the Dominican Republic. It is a regional institution that combines activities in traditional production-oriented sciences with modern conservation and environmental management.

IUCN normally builds on and helps to strengthen existing initiatives; its philosophy is one of long-term commitment and continuous follow-up. The monies provided by IUCN are usua!ly "seed or bridge funding" which helps promising ideas and successful projects to grow at the appropriate speed. The projects tend to adopt a step-by-step approach, linking a series of projects to address a specific conservation for development problem. IUCN's Office for Central America, based in San José, Costa Rica, provides coordination and liaison services to regional, national, and local organizations and institutions. It can often serve as a convenient buffer and focal point in contacts between donors and local recipients. Most important, the office helps initiate and coordinate projects, primarily:

- *projects that demonstrate the link between conservation and development*
- *environmental impact assessments*
- *institution building*
- *human resources development*

This paper reviews the development and implementation of these projects in Central America. The projects are summarized in Figure 1 and Table 1.

2. DEMONSTRATION PROJECTS: IMPLEMENTING THE WORLD CONSERVATION STRATEGY

One of the most effective means of gaining wider support for conservation and sustainable development is the implementation of demonstration projects. These action-oriented projects illustrate the importance of wise resource utilization to decision-makers, the rural population and the general public. They also function as open-air laboratories for new technologies and management techniques, or as outdoor activities, and are used to train Central American resource managers.

The demostration projects can be divided into: (a) strategic planning and implementation and (b) small-scale eco-development projects that support the planning projects.

The strategic planning and implementation projects are based on the World Conservation Strategy and are carried out both nationally and sub-nationally. Interinstitutional and multidisciplinary planning and implementation are emphasized. The focus is long-term sustainable development, and both "preservation" and "exploitation" projects form part of these strategies. It is hoped that all seven Central American countries will eventually develop their own National Conservation Strategies for Sustainable Development.

The eco-development projects complement the overall planning efforts by demostrating how conservation and natural resource management can guarantee sustainable long-term benefits and often attain substantial increases in short-term profits. They provide immediate results and maintain interest in larger planning schemes while experimenting with new solutions to relevant, local development problems.

Since foreign scientists and development officers are seldom able fully to take into account the traditions and attitudes of local inhabitants and the characteristics of the local environment, IUCN works almost exclusively with Latin American experts in the region. The demonstration projects are carried out in association with IUCN members when possible. IUCN provides technical and logistical support, oversees and manages the projects when needed and helps raise the necessary funds.

2.1 National Demonstration Projects

Panama

Sustainable Development Strategy for tropical forests, coral reefs, coastal wetlands and lagoon systems of the Bocas del Toro Province

Together with IUCN, the Government of Panama is designing a sustainable development strategy for the Province of Bocas del Toro, one of the largest untouched tracts of tropical forest expanses in Central America. The Province includes the Panamanian portion of La Amistad International Park - a major forest reserve - as well as internationally important coral reefs, coastal wetlands and lagoon systems.

During recent years the pressure for "development" of the province's resources has increased substantially. Bocas del Toro contains most of Panama's remaining hydroelectric potential. In addition, it is likely to experience rapid population growth in coming years, as a result of recently constructed roads connecting the province to the rest of the country and Costa Rica, and a port where an oil pipeline has its Atlantic terminal.

Other issues influencing the development of the province include an acrimonious dispute between indigenous communities and the central government over land rights, and dependence on one cash crop for over a century - the banana.

In 1984, IUCN representatives met with Panamanian government officials and local politicians, and undertook the initial planning phase for the sustainable development of the region. The resulting proposal achieved wide support in Panama and the main working phase of the Strategy has been underway since mid-1987. The Provincial

Strategy team encourages local participation and leadership. Several basic studies needed to make decisions about appropriate development schemes have been negotiated and some initiated, even though the political situation in Panama has made work difficult. The Strategy team is cooperating with several national agencies, institutions and NGOs as well as with international organizations.

The main working mode during the first years is to implement several small projects in order to demonstrate the meaning of "conservation for sustainable development" so that the options for change are understood by the the local people, and their views are taken into account in determining the direction taken by development. Currently, the entire project is being reviewed to determine how the different technical programmes of IUCN might best work together within a consolidated framework.

Costa Rica

National Conservation Strategy for Sustainable Development

In 1986, the Costa Rican Government requested assistance with the development of a National Conservation Strategy (NSC) in order to integrate sound management of the country's resources with its long-term development plans. The international environmental community responded quickly and, by early 1987, the Strategy process was initiated. The recently created Ministry of Natural Resources, Energy and Mines plays the leading role in the development of the Strategy.

Appoximately 90 high-level professionals have been involved in the preparation of sector strategies based on a general framework that guarantees compatibility. Several intensive workshops have been held to synthesize each sector strategy. The final reports have been reviewed by experts. The sectors will be synthesized into one strategic document, which will then be subject to a thorough review by private enterprise, government agencies and the general public.

The findings - which will form the basis for a new and different development style for the nation - have been discussed at a series of national hearings. The President and most Ministers, as well as representatives from the opposition parties, participated. Several specific programs and projects in need of funding were presented within a long-term conservation for development framework that aims to guarantee their success.

Another component of the Strategy process is a massive media campaign to inform the public about the relationship between their every-day use of resources and the future development of the country, i.e. pollution issues, use of pesticides, and effects of deforestation. A similar awareness campaign will also be necessary within government and NGOs in order to allocate responsibilities and ensure the continuation of interinstitutional and multidisciplinary collaboration.

Economic Evaluation of the Rio Macho Forest Reserve

The Rio Macho Forest Reserve, which forms part of the Talamanca Biosphere

Reserve, is one of the most successful and important of the wide network of forest reserves in Costa Rica. Approximately half the drinking water for the heavily populated Central Valley originates there and the Reserve makes a substantial contribution to hydroelectric power generation. The area also receives great attention owing to a growing interest in nature tourism.

Rio Macho's contribution to the national economy was analyzed in a study carried out by Costa Rican economists, foresters and sociologists. The multidisciplinary team identified land use alternatives, determined costs and benefits associated with each, and compared the net benefits of development activities that do not eliminate forest cover with the net benefits of activities that do.

The results of the study, when throughly reviewed and evaluated, will serve several purposes: *First*, it appears that the study will demonstrate clearly that conservation is the most "economic" use of the forest's resources, given the high values of the ecological services it provides to society. It offers an all-too-rare documented demonstration that conservation does not necessarily imply foregoing the use of resources, but rather making a different (in this case equally productive) use of it: *Second*, the methodology developed for this project could constitute an important contribution to management of forest reserves.

Sustainable Use of the Green Iguana in Talamanca Communities

The Green Iguana is an important source of protein (meat and eggs) and income (skins) for the indigenous and campesino (subsistence farmers) communities of the Talamanca region in south-eastern Costa Rica. The popularity of the reptile, combined with the destruction of its rainforest habitat, has put it on the endangered species list.

In the last few years, researchers have worked with campesinos in Panama to develop a successful breeding programme for the iguanas in their communities. Inspired by this and by earlier tree and plant nursery projects, community leaders in the Talamanca region and the Costa Rican eco-development group (ANAI) approached IUCN with the ideas of developing a similar project in their communities.

The project started in late 1987 in two locations, the Cocles Indigenous Reserve and the community of San Miguel. At each site a reproductive colony has been established, with all contructions donated by the communities. An regional educational campaign has been launched by community leaders and the iguana habitat is being reforested with economically valuable trees. The iguanas will eventually be released directly into the wild or into the stewardship of selected families.

Both sites are located in the buffer zone surrounding the Gandoca-Manzanillo Wildlife Refuge on the Caribbean Coast. It is hoped that the project will provide a viable development alternative, demonstrate the importance of the rainforest habitat for the economic well-being of the local population and help alleviate pressures on the refuge.

In 1989, the research centre was moved from Panama to Costa Rica and will broaden its focus to include training, extension and demonstration projects on village-level iguana management.

Nicaragua

Heroes and Martyrs of Veracruz; Rural Integrated Redevelopment of the Nagrandana Plain

IUCN has cooperated with Nicaragua's Institute for Natural Resources and Environment on several activities that have resulted in one large-scale integrated natural resource management and rural development project: Heroes and Martyrs of Veracruz. It covers 2,000 square kilometers and is bordered to the north-east by the Maribios volcanic range and by the Pacific coast to the west.

The area's 336,000 inhabitants have seen the Nagrandana plain between the coastal mangrove forest and the mountains go from being the most fertile agricultural area of the country to an area devastated by poor resource management and monoculture plantations. The impacts have been many: valuable soil and water resources have been lost, the principal routes connecting the area with the capital city have been completely eroded, the coastal ecosystems has been severely disturbed, the supply of wood and other forest products for household consumption has decreased, among others.

The Heroes and Martyrs project brings together national, regional and community organizations in an effort to develop an integrated approach to the whole set of problems. Erosion control, water harvesting, reforestation, wildlands management, environmental education and control of forest fires are but a few of many components. The active participation by local cooperatives is emphasized as an objective in itself, while it is realized that basic research is also needed.

The value of the annual production of the region's three major products - cotton, sugar cane and bananas - was estimated at more than US$180 million in 1986 in a country with severe economic problems. The project, though large and costly, accounts for only 1 per cent of the annual gross value of that produce, while promising to augment the rate of return and guarantee the long-term sustainability of the regional production system.

In 1986, work began on a pilot area within the overall project - Pikin Guerrero watershed. An integrated approach has been taken to rehabilitation of the severely degraded watershed, from soil erosion and gully control to tree-planting, water management and pest control. The activities are implemented by the village cooperatives with assistance and material support from IUCN and Nicaraguan governmental agencies. IUCN is also developing a management scheme, in cooperation with a rural development programme (Chinorte), for the steep slopes of the Maribios range and for the mangrove forests of the estuary, Estero Real.

Honduras

Potable Water for Tegucigalpa

During the dry season, more than one third of Tegucigalpa's 650,000 residents are without direct access to water. Instead, many of them are forced to buy poor quality supplies from water hawkers. This is in part because the capital city is badly located, but more important, factors such as deforestation and pollution of river basin areas as well as rapid population growth combined with a lack of urban planning have brought the city to its present straits.

In addressing this complex problem, it was found that La Tigra National Park plays an essential role in the city's water supply. This 7,000 ha cloud-forest reserve, located 11 km from Tegucigalpa, has few resources available to implement its management plan and is guarded by only a handful of people. Yet, the treatment cost for the water from La Tigra is only 1/23 of that of the second best alternative.

At a workshop held in Tegucigalpa in April 1986, these and several other scientific findings were presented to more than 100 participants. IUCN was then asked to assist the Honduran government in drawing up a detailed strategy for the potable water situation and responded with a follow-up seminar in April 1987. An interinstitutional approach to the problems identified at the workshop was prepared by the 20 participating government agencies and NGOs.

It was decided that clear areas of responsibility needed to be delineated and joint projects carried out such as forest exploitation control, protection of watersheds, reorganization of land use patterns, migration control through economic incentive schemes, socio-economic studies in the watersheds, and environmental education programs.

It was also determined that overexploitation of areas around La Tigra National Park as well as privately held inholdings (32 per cent of the park) would cause serious deterioration of the water supply if allowed to continue. An interinstitutional team of resource managers agreed that in order to protect La Tigra, a buffer zone of 14,490 ha should be created around the 7,750 ha park.

Studies of the local communities and land use practicies in the area will allow the development of educational and extension programs aimed at improving land use patterns and controlling deforestation and further degradation of La Tigra. Key demonstration areas are to be selected by an interinstitutional team and a six-year master plan for the zone with short-term action plans will be developed.

El Salvador

National System of Protected Areas

El Salvador has the greatest population density and the highest level of environmental destruction in Central America. No more than 12 per cent of the country is covered by trees and only 3 per cent can be termed wildlands. In 1980, the impoverished rural population was given land from haciendas of more than 500 ha as part of the first phase of governmental agrarian reform; however, they were also required to pay the Government for the land over the ensuing 20 years.

Some of these haciendas included large areas of forest not suitable for agriculture, which the peasant cooperatives were not interested in acquiring. These areas have now been placed in reserve by the Goverment. IUCN is assisting a governmental agency with the creation of a system of protected areas, working to identify appropriate sustainable development approaches for these areas and their surroundings that will benefit the local population. Policies and legal instruments capable of supporting the system of protected areas will also be developed.

The approach taken by the project is highly innovative. The governmental agency does not have the capacity to take on the network of over 50 new areas for which it has aquired responsibility. The project will therefore examine how management responsibility might be shared by local governments, municipalities, peasant cooperatives, and NGOs.

The general approach is based on the World Conservation Strategy and is expected to lead to the initiation of a National Conservation Strategy. It is interinstitutional and multi-disciplinary and includes representatives of most government institutions and NGOs.

Guatemala

Regional Conservation Strategy for Peten

In October 1987, the Guatemalan Government formally requested IUCN's support and assistance in preparing a National Conservation Strategy. After several preparatory discussions it was decided that the first step should be the preparation of a Conservation Strategy for the Peten region.

Peten covers 35,858 square kilometers or 32.9 per cent of Guatemala and since 1959 has been governed by the National Agency for Promotion and Economic Development of Peten (FYDEP, directly under the Defence Ministry). In a recent analysis of FYDEP's last five years, it was found that the income generated had decreased substantially while the population had increased by aproximately 10 per cent annually. This performance was explained by the central government as the result of a deficient management of natural resources and a lack of clear strategic planning for the use and conservation of the region's resources.

The Guatemalan Government recently created a Regional Development Council for the Peten and gave it authority to make decisions about the future of the region. The Regional Council - whose members represent both the public and private sectors - has prepared a first draft of a "Regional Conservation Strategy for Sustainable Development" within which all development activities are to be integrated. The strategy was presented at a national workshop in July 1988, and was accepted. Similar strategies for Guatemala's eight other regions are being prepared.

Belize

Coastal Environmental Planning: The Belize Barrier Reef Complex

The Belize Reef Complex is second only to Australia's Great Barrier Reef in size and complexity. It is on average 30 km wide over the entire 250 km from the northern tip by Mexico to the southern tip by Honduras. Its assortment of ecosystems is of great global, regional and national importance, consisting of the barrier reef, patch reefs, fringing reefs, continental shelf atolls, sheltered lagoons, mangrove cays and extensive seagrass meadows.

The reef has a great fisheries and nature-based tourism potential and forms a natural barrier to frequent hurricanes. Until recently the use of the reef has been so limited that no noticeable environmental destruction has ocurred. However, tourism is increasing rapidly; mangrove swamps on the mainland are being cleared to provide landfill for construction or space for shrimp farms; seagrass beds are being dredged and deforestation is beginning to generate soil, pesticide and fertilizer run-off and coastal sedimentation, owing to a lack of regulation or a limited ability to enforce existing legislation.

Coastal management with a "conservation for development" approach is needed and IUCN has offered to collaborate with the Belize Goverment and the Belize Audubon Society to develop a management plan for the reef complex. The plan calls for increased cooperation with international organizations operating in the region, as well as with Belize authorities.

3. REGIONAL DEMONSTRATION PROJECTS

3.1 Conservation for Sustainable Development in Central America

CATIE and IUCN have developed an integrated approach combining production and conservation sciences for sustainable development and now hope to translate it into practical programs for action in Central America. A network of governments institutions, universities and NGOs has been established, through which a series of pilot projects demonstrating the benefits of sustainable resource management will be carried out.

A three-year programme has now been launched that will include demonstration projects such as:

- *zoning of coastal wetlands in Panama*
- *investigation of sustainable forest resources*
- *management in indigenous communities in Costa Rica*

- *management of forest ecosystems and wetlands in Nicaragua*
- *development of a system of protected areas for Peten in Guatemala*

The projects will allow institutional strengthening and professional exchanges within the Central American region. The initiation of a long-term conservation for sustainable development programme represents a significant step towards sustainable development in Central America.

Trinational Biosphere Reserve: Trifinio

Several years ago, the governments of El Salvador, Honduras and Guatemala proposed the creation of a trinational park on their borders. Over time this was modified into a large-scale integrated rural development scheme, developed under the auspices of the Organization of American States and the Inter-American Institute for Agricultural Cooperation (IICA). IUCN is providing assistance to the three countries for the design of an approach to conservation and development in the newly created "Fraternidad" biosphere reserve at the centre of the project area.

In November 1987, the three Vice Presidents met and signed a proposal to UNESCO for the establishment of a Biosphere Reserve in the region. Later the same year IUCN was formally asked by the three governments to help coordinate the preparation and implementation of a management plan for the proposed reserve and to take responsibility for the environmental education component.

An IUCN project design mission visited Trifinio in October 1988 to work out the details of the protected area plan and environmental programme.

Much discussion has also taken place on the possibilities of developing a coordinated series of projects and protected areas on either side of the Suan Juan river, which forms the border between Nicaragua and Costa Rica.

4. ENVIRONMENTAL IMPACT ASSESSMENT

Environmental impacts assessments (EIA) cover a wide range of areas. IUCN's usual role is to provide a technical framework and financial assistance for a consultant to evaluate the project. The assessment is always done in cooperation with the local community. The following preassessment missions have been undertaken or are subject to final confirmation:

Costa Rica

Impacts of Road Constructions

A preliminary EIA of the effects on rural communities of a road proposed along the Caribbean coast of Costa Rica was undertaken. The road would have traversed an important wildlife refuge. As a result of the IUCN report carried out in partnership with Costa Rican based NGOs, the government abandoned the road project and declared the area a national wildlife refuge.

Another assessment further north along the same coast analyzed the impacts of an ongoing extension of the national road network, bringing settlers up to the edges of Tortuguero National Park and the Barra del Colorado Wildlife Refuge. The report recommended actions to be taken to mitigate the negative impacts of the expansion.

Impacts of a Large-Scale Irrigation Project

An EIA was carried out on behalf of the Costa Rican Government and aimed at integrating conservation into a major irrigation project in the Guanacaste Province. The work played a substantial role in their proposal to the funding agency, the InterAmerican Development Bank, for a comprehensive environmental program as one component of the project. Another major result of the study was an increased awareness in the implementing engineering agency of the environmental dimensions of the irrigation project. The cooperation between the Costa Rican government agencies and IUCN in assuring the long-term success of the irrigation project continues, now at the request of the engineering agency.

Sedimentation Impacts of Banana Plantations

A preassessment of the impacts on coastal ecosystems of sedimentation from banana plantations in Talamanca, Costa Rica, and Bocas del Toro, Panama will be performed. The primary objective of the study is to suggest alternatives or mitigation measures, if such are necessary and available.

Guatemala

Impact of a Pulp and Paper Mill

An assessment was undertaken of the environmental impact of a proposed pulp and paper plant in the province of El Progreso, one of the largest economic development projects ever planned in Guatemala. The study looked at the potential environmental impacts and specified measures to reduce the most serious effects. As a disinterested, non-political, scientific organization, IUCN was able to enter into an intensely partisan and highly emotional project and provide advice that was acceptable to all parties. Following the publication of the EIA the Guatemalan government denied permission to the plant to begin operations until environmental safeguards were in place.

Belize

Introducing Environmental Impact Assessment Procedures

The design of a minimum environmental assessment capability for Belize is being discussed, in particular in connection with the operations of the Belize Export and Investment Promotion Unit, whose responsibility includes review of all development proposals, both in the private and public sectors.

5. INSTITUTION BUILDING FOR ENVIRONMENTAL PLANNING AND ACTION

The purpose of institution building is to develop a self-sustaining regional conservation for sustainable development programme and help build a capacity to promote and lead activities in this field, both nationally and regionally. Small promising NGOs that take on the complex issues of conservation, natural resources management and economic development are provided with modest support in their initial phases. More established organizations and institutions are given specific projects.

6. DEVELOPMENT OF HUMAN RESOURCES

In order to help the region become completely self-sufficient in the field of natural resources management, IUCN is facilitating training opportunities and professional interaction on a national and regional basis. Exchanges of specialists between different developing countries and regions are promoted. The objective is to build institutional and personal links between conservationists and natural resource managers throughout the developing world and thereby help speed up the development of institutional capability for handling conservation-related development issues.

7. FUTURE DIRECTIONS FOR IUCN IN CENTRAL AMERICA

In the last few years, the Central American environment has seen many changes - for the better and for the worse.

Deforestation continues undabated: one estimate finds that the region is losing

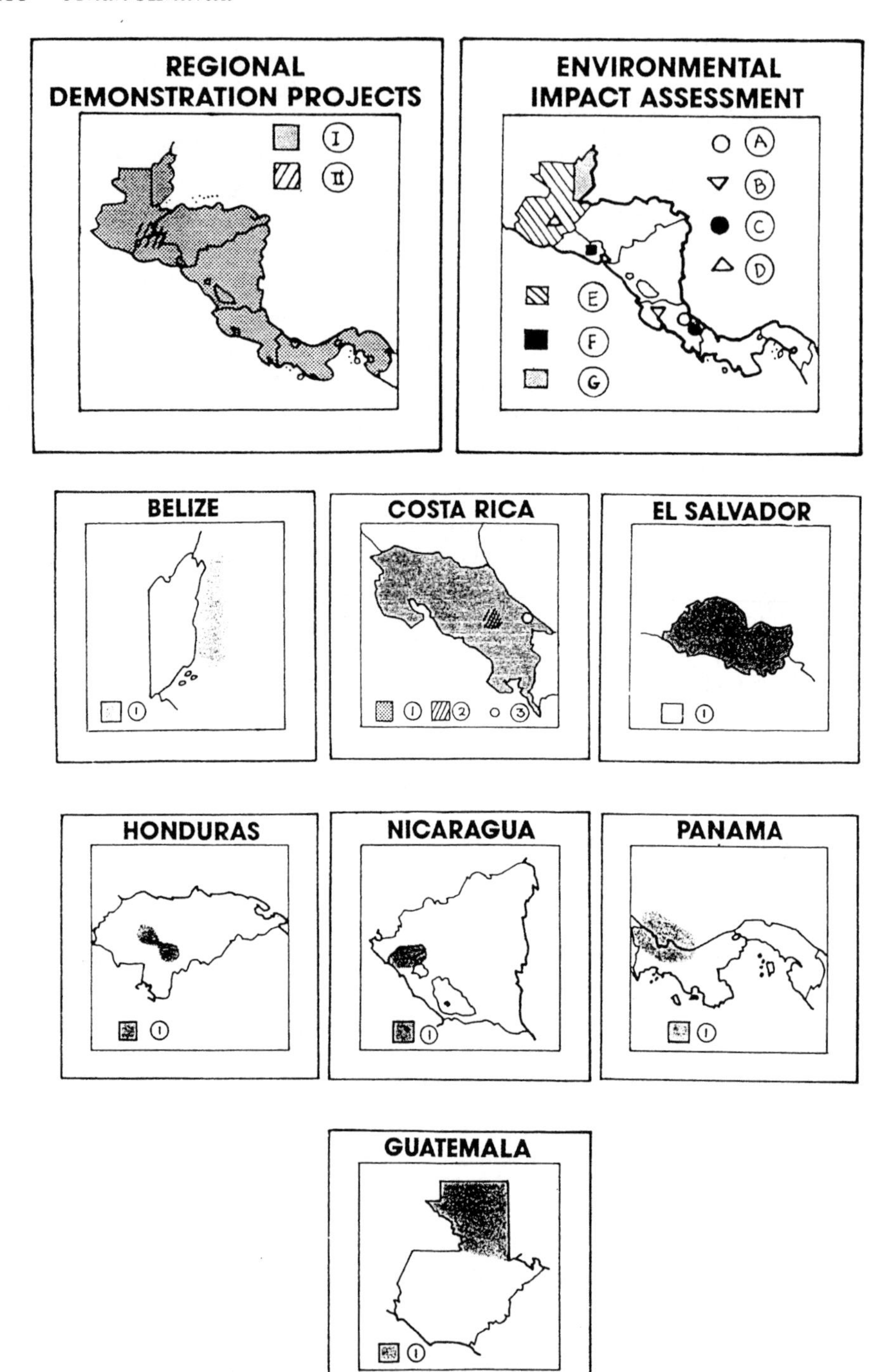

Figure 1. Maps of Central America, by region and by country, showing the location of the projects summarized in Table 1.

Table 1. Conservation for Development in Central America: IUCN. Numbers and letters refer to the maps in Figure 1.

Title	Location	Duration	Sponsor
REGIONAL			
I. Conservation for Sustainable Development in Central America	Central America	1988-	NORAD/SIDA
II. Trinational Biosphere Reserve: Trifinio	Honduras Guatemala El Salvador	1987-	NORAD/SIDA (phase 1)
- Assistance to CATIE	Costa Rica	1985-	CIDA
- Assistance to Government Ministers, and NGOs	Central America	1985	Various
- Support to Regional Conferences	Central America	1987-1988	NORAD/CIDA
- Regional Working Group of Conservation Experts	Central America	1987-	NORAD
- Internship/Fellowship Programme	Switzerland	1989	SIDA
BELIZE			
1. Coastal Environmental Planning: Barrier Reef Complex	Belize		
G. Introducing Environmental Impact Procedures	Belize		NORAD/CIDA
- South-South Exchange	Belize-Australia		CIDA
COSTA RICA			
1. National Conservation Strategy for Sustainable Development	Costa Rica	1986-88	SIDA
2. Economic Evaluation of the Rio Macho Forest Reserve	Rio Macho	1986-87	NORAD
3. Sustainable Use of Green Iguanas	Talamanca	1987-	NORAD
A. Environmental Impacts of Road Construction	Talamanca	1986-	NORAD/SIDA
B. Enviromental Impacts of a Large-Scale Irrigation Project	Guanacaste	1986-	NORAD/SIDA

Table 1, continued

Title	Location	Duration	Sponsor
C. Sedimentation Impacts of Banana Plantations	Talamanca	1988-	NORAD/SIDA
- South-South Exchange	Costa Rica- Thailand Southeast Asia India-Pakistan	 1985- 1986 1988	 CIDA NORAD
EL SALVADOR			
1. National System of Protected Areas	El Salvador	1988-	
F. Water Levels in the Jocotal Lagoon	Jocotal Lagoon		
GUATEMALA			
1. Regional Conservation Strategy for Peten	Department of Peten	1988	NORAD/SIDA (phase 1)
D. Environmental Impact of a Pulp and Paper Mill	El Progreso	1986-	NORAD/SIDA
E. Environmental Impacts of Proposed Medfly Eradiction Campaign	Guatemala	1988-	NORAD/SIDA
HONDURAS			
1. Potable Water for Tegucigalpa	Tegucigalpa	1986-	NORAD/CIDA
NICARAGUA			
1. Heroes and Martyrs of Veracruz	Region II	1986-	NORAD/SIDA
- South-South Exchange	Guatemala-Costa Rica	1986-	CIDA
PANAMA			
1. Sustainable Development Strategy for Bocas del Toro Province	Bocas del Toro	1985 1986-87	WWF (phase1) NORAD (phase II)

CIDA = Canadian International Development Agency
NORAD = Norwegian Agency for International Development
SIDA = Swedish International Development Authority
WWF = World Wildlife Fund for Nature

2.9 per cent of its forest cover every year. Pesticide poisonings are on the increase, precious coastal resources are disappering and freshwater supplies have become dangerously low.

However, the governments of the isthmus are now firmly behind sustainable development efforts (although the level of commitment varies), and the number of functioning environmental NGOs is increasing rapidly. Every nation now has an environmental ministry or commission. Regional initiatives are also becoming popular - the recently formed Central American Parliament will most likely include an environmental commission, while governments with shared borders are working to develop international parks.

Much of this change has been caused by the realization that countries saddled with large populations, overwhelming external debt and a history of environmental exploitation will benefit economically from policies that encourage the sustainable use of natural resources and ecosystems. Wise management of their ecosystems gives Central America the possibility for a brighter future.

Through building partnerships with local institutions and providing sorely needed technical and financial assistance, IUCN has been able to make substantial contributions to the Central Americans'struggle to work with their environment - and not against it.

The international financial support for this work - specifically from the governments of Norway, Sweden and Canada, as well as a range of others, has allowed IUCN and its partners to build up a programme based on the highest priorities for conservation and development action (Table 1).

This support is helping the programme to grow and evolve. Several initiatives are currently in the planning stage and many will soon be underway. One of these is a major regional programme on use of wild resources by local communities - exploring the potential for poor rural communities to improve their standard of living through sustainable use of the full range of wild resources that surround them, whether medical plants, ranched or wild-caught animals, honey or orchids.

The work concerning wetland issues will also be greatly expanded, and a regional wetlands advisor has recently been recruited.

At the national level, new initiatives are being designed such as strengthening the green belt around the capital of Guatemala, and implementing an ecodevelopment project with the fishing communities of Livingstone on the Caribbean coast. In Costa Rica, IUCN has been advising Sweden on programming of new funding for conservation.

Major population growth, overexploitation of natural resources and pollution of the environment are rampant in Central America. Nevertheless, changes are underway as it becomes economic for countries to pursue a path of sustainable development. It is hoped that the work described in this paper will make changes easier to plan and easier to implement, lighting the way for a brighter future in Central America.

The regional program for Central America is evolving and will adjust to the needs of the region and eventually become obsolete when the independent institutional capability to handle conservation-related sustainable development issues has been firmly established.

ACKNOWLEDGEMENTS

This paper was originally written by Johan Åshuvud in his capacity as IUCN project representative for Central America. The editors thank the IUCN for allowing us to include this modified version of Åshuvud and Whelan (1988), updated in 1989 by Tensie Whelan for this book. For more information concerning the projects please contact Mark Halle, IUCN, Gland, Switzerland.

REFERENCES

Åshuvud, J. and Whelan, T. 1988. IUCN Regional Programme for Central America. IUCN Regional Office for Central America, San José, Costa Rica. 16 pp.

World Conservation Strategy. 1980. IUCN, Gland, Switzerland.

Part IV

SYNTHESIS

Linking the Natural Environment and the Economy; Essays from the Eco-Eco Group,
Carl Folke and Tomas Kåberger (editors)
Second Edition
1992. Kluwer Academic Publishers

CHAPTER 14

Recent Trends in Linking the Natural Environment and the Economy

by

Carl Folke and Tomas Kåberger

Beijer International Institute of Ecological Economics
The Royal Swedish Academy of Sciences
Box 50005, S-104 05 Stockholm, Sweden
and Department of Systems Ecology, Stockholm University

Physical Resource Theory Group
Chalmers University of Technology
S-412 96 Gothenburg, Sweden

The final chapter focus on some of the recent trends in linking the natural environment and the economy which have evolved, particularly in the interface of ecological and economic sciences. We have divided the chapter into two parts. The first part provides a brief overview of some major perspectives on environment in development, and relates the previous book chapters to these perspectives. The second part emphasizes the significance of "maintaining natural capital", a concept on which there is a rapid merging of ecological and economic thinking. We relate this concept to a discussion on sustainability, based on recent attempts to define and clarify what is a sustainable society.

1. INTRODUCTION

In the last decade there has been a rapid development of research aimed at linking the natural environment and the economy. The increased interest and the urgency of approaching a sustainable relation between the socio-economic system and the natural environment involve a wide range of disciplines, from basic natural sciences such as physics and chemistry to those dealing with social structures, economic and juridical considerations as well as cultural and ethical perspectives.

It would be a too formidable task to try to cover all these recent research trends. To a large extent constrained by our own pre-analytic visions, we will instead concentrate on some of the recent ideas and important perspectives on environmental management that have evolved, particularly in the interface between ecological and economic sciences. We have divided this final chapter into two parts. The first part provides a brief overview of some major perspectives on environment and development, and relates the previous chapters of this book to these perspectives. The second part emphasizes a concept on which there is rapid merging of ecological and economic thinking; the significance of "maintaining and investing in natural capital." We relate this concept to a discussion on sustainability based on recent attempts to define and clarify what characterizes a sustainable society. At the end of the chapter we have made a list of some relevant books for those interested in reading more about the importance of linking the natural environment and the economy.

2. ECOLOGICAL AND ECONOMIC PERSPECTIVES ON ENVIRONMENT AND DEVELOPMENT

The disciplines of ecology and economics have many things in common. Both of them attempt to understand and predict the behavior of complex interconnected systems in which both individual behavior and large-scale flows of energy and material are important. The two disciplines use similar quantitative tools such as input-output analysis, simulation, and maximizing calculus. They share similar concepts such as competition and specialization, and they are both concerned with open systems having one major external input, energy (or exergy) that is not reusable - sunlight for ecosystems and mainly non-renewable fuels for economic systems. Both are structured by the decisions of individuals, who function in the context of hierarchies of group organizations, interacting with their environment (Bernstein, 1981; Harris, 1985).

2.1 A Synthesis of the Essays in this book

Whether rooted in economics, ecology, or other sciences, the essays of the Eco-Eco Group have in common that they deal with environmental and socio-economic issues in an integrated manner. They do, however, represent a diversity of approaches to such interrelated issues. In Figure 1 we have developed a simplified conceptual model to illustrate where this diverse assembly of essays belongs. The lower part of the model represents social sciences and the upper part natural sciences. The left half concerns those approaches using several evaluation criteria for analyzing the interactions between ecological and economic systems, and the right half those using a common denominator for this evaluation, such as money or energy. There are several perspectives within each box as well as intermediate forms, both vertical, horizontal, and cross-sectional, indicated by the arrows in the model. Especially during the 1980's, the boundaries between the different boxes have become less clearly defined, which implies that there is an ongoing melding and integration of various approaches, concepts and methods, but also an increased diversification of interfaces and contexts aiming at linking the natural environment and the economy. Crucial references to the major perspectives in Figure 1 are found in the essays of this book and also among the selection of recent books at the end of this chapter.

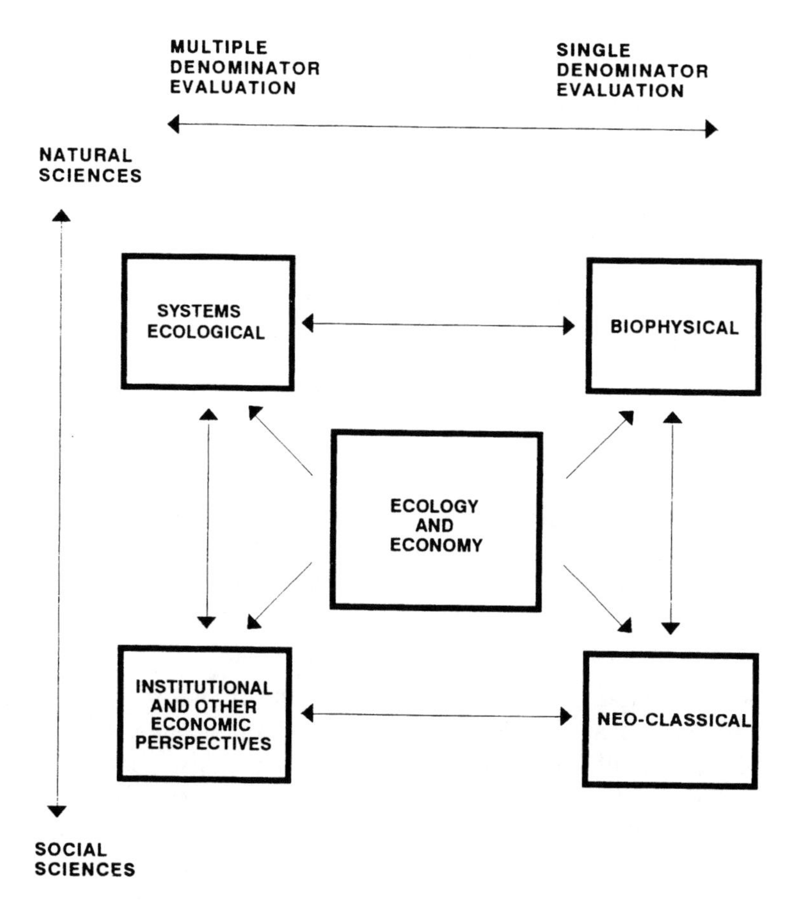

Figure 1. A simplified conceptual model of ecological and economic perspectives and approaches to environmental issues, applicable to the essays of this book (see text).

Returning to the chapters of this book, the first, by *Svedin*, providing us with contextual features of the economy-ecology interface, encompasses all the four boxes in Figure 1, and also non-economic/ecological approaches not explicitly covered by our conceptual model. Svedin emphasizes that there are several economies as well as ecologies, and that this overlapping pluralistic research area of intersecting connections is highly dependent on the contextual framework. This context, in turn, has many facets connected with basic value issues and related to their socio-economic and cultural frameworks. He argues that the contextual nature of concepts, ideas about relations and the societal embedding of the problems are at the heart of both environmental problems and solutions. If we were forced to give his essay a position, we would tentatively place it at the left half of the model, with a slight bias towards the lower part.

Chapter 2 by *Söderbaum*, belongs to the Institutional School of economic thinking (see also Figure 2), and it emphasizes the significance of explicitly stating the valuational premises of one's research, and wherever possible considering several valuational viewpoints rather than just one. He emphasizes the use of concepts such as actor, role, network, power, institutions, values, and ethics for dealing with environmental issues. He argues in favor of ecodevelopment and states that a research agenda for ecodevelopment has to face all aspects of environmental issues, such as the problem with different paradigms, the problem of knowledge and lack of knowledge, the problem of professional roles, lifestyles, business concepts and development concepts at large, as well as the problem of power relations and institutional design.

The chapters by Bojö, Andersson, Andréasson-Gren and Hasund take their economic standpoint in the neoclassical perspective.

Bojö (Chapter 3) discusses some reasons why markets and policies fail to adequately address environmental problems. In a comprehensive way he summarizes economic methods that can assist the analysis of environmental costs and benefits, focusing on project level analysis using cost-benefit analysis. Particular attention is devoted to economic valuation of environmental impacts.

Andersson (Chapter 13) analyzes the role of multinational enterprises in the production and location of pollution-intensive activities in developing countries. He examines host country competition for investment projects, using a game theory framework, and concludes that the causes of rapid environmental degradation are to be found not in the interaction between multinational enterprises and competing host countries, but in imperfections in the international capital markets, in access to information and in government failure.

In Chapter 9, *Andréasson-Gren* performs a straightforward cost-effectiveness analysis of different measures for nitrogen source reduction to a eutrophicated bay in Sweden. Among the measures analyzed are the use of fertilizers, treatment of manure, and other agricultural adjustments, restoration of wetlands, sewage treatment, and traffic emissions. From the cost-effectiveness calculations she concludes that agriculture must take responsibility for the largest part of the total nitrogen reduction to the bay. Interestingly, using the free work of nature by means of restored wetland ecosystems is the least costly measure, amounting to only one third of the marginal cost of the second cheapest alternative.

In Chapter 7, *Hasund* explicitly links the environment and the economy by generating both biophysical and economic measures for evaluating arable land as a production resource, and concludes that the barley-equivalents and land rents methods, which he has developed, each provide a single, cardinal measure appropriate for comparing arable land of different quality. Hasund also applies a computerized model of Swedish arable land resources to study the impacts of various resource influencing factors such as air pollution, subsoil compaction and urban expansion. In particular, the last two neoclassical essays expand their scope of research into the upper part of Figure 1, by explicitly recognizing the socio-economic significance of environmental functions and ecosystem services.

The socio-economic significance of the life-supporting natural environment is further emphasized in Chapter 5 by *Folke*. This ecological economic essay, founded on systems ecology, is concerned with humans with their environment, the man-in-nature view as opposed to the man-against-nature view. The role of the environment in supporting the economy is identified, and related to growth and sustainability issues. This essay concludes that a major challenge in the effort to approach sustainability is to enable the agents of the human economy to fit the socio-economic systems into biogeochemical processes so as to maintain the life-supporting environment on which human societies depend.

Three case studies investigate the interdependence between the life-supporting environment and the human economy.

The first by *Jansson* (Chapter 6) is a regional ecological economic study of urban/environmental landscape changes on a Swedish island. The pioneering work presented in this essay illustrates interrelations between human settlements and the support provided by the work of the natural environment over a period of almost three centuries. Her study explores the transformations and productivity changes of terrestrial ecosystems and shows that during this period the potential work contribution of these ecosystems to the regional economy has been reduced as a consequence of human activity. Her essay also reveals a pulsing pattern of prosperous growth and decay in the development of human societies, to a large extent related to the use and imports of energy, resources, and ecosystem support, as well as various cultural influences.

The second study (*Folke*, Chapter 8) is an attempt to evaluate the lost life support of a degraded Swedish wetland system and to estimate the money and industrial energy invested in technical solutions to replace the wetland's production of environmental goods and services. These costs were estimated to be within the same order of magnitude as the loss of the free work of the unexploited wetland. The investigation illustrates the fact that exploitation of natural ecosystems believed to be of limited use to man often leads to high costs for society, one important reason being that the ecosystem support is not recognized or fully acknowledged until the environmental damage has already been done.

In the third study on the significance of the life-supporting natural environment, *Hammer* (Chapter 10) compares trade balances for fish products in economic and ecological terms. She shows that although in money terms the Swedish imports are 2.5 times greater than the exports, in ecological terms they are about 7.5 times the exports, meaning that Sweden is receiving natural resources and the support from other ecosystems at a lower price than that at which Sweden is selling its own environmental goods and services. By using biophysical estimates, Hammer shows that socio-economic activities are much more dependent on ecological support than is indicated in standard economic valuation, where generally only man-made goods and services are considered.

These three ecological-economic analyses specifically belong to the systems ecology and biophysical boxes of our conceptual model.

Another essay which we place in the biophysical box is the one by *Kåberger* (Chapter 4). He analyzes and discusses the pros and cons of different measures which use energy as a common denominator in the study of human economies, the natural environment and their interrelations. He distinguishes between analyses using energy content and those using energy cost of production. Kåberger concludes that using various energy measures in the analysis of interacting economic and ecological systems is very useful, although perhaps unnecessary, and that it is certainly not sufficient in order to design an environmentally acceptable economic theory or policy.

In Chapter 11, *Christiansson* brings down to earth the issue of how socio-economic activities and the natural environment interrelate, by investigating the use and impacts of chemical pesticides in smallholder agriculture in Kenya. The survey describes and discusses impacts on human health as well as on the local environment. The study reveals that agricultural pesticides are widely used, and that considerably more is used on cash crops than on subsistence crops. Christiansson found that many pesticides forbidden in the industrialized world are widely used in the study area, and that over 50 per cent of the farmers interviewed had experienced health problems probably related to the handling of the pesticides. This essay conforms to the left part of our conceptual model (Figur 1).

The last essay, by *Åshuvud* (Chapter 13), to whom this book is dedicated, describes an impressive variety of conservation for development projects in Central America. Based on the World Conservation Strategy, a major objective of these projects is to translate the concepts of natural resource management, conservation and sustainable development into practical programmes for action through local institutions. Conservation is not seen as a separate sector of development, but rather as a concept unifying different sectors of society as well as governmental and non-governmental activities. Åshuvud stresses the need for ecodevelopment (see Figure 3), and concludes that the development potential depends in part on the capacity of local institutions to successfully implement sustainable development projects. Wise management of their ecosystems would give Central America the possibility for a brighter future. This essay encompasses the major part of our conceptual model.

From this brief review of the essays and their scientific pre-analytic visions we will continue by summarizing what are, in our opinion, some important contributions for clarifying the various approaches evolving on the role of environment in development.

2.2 Evolution of Paradigms in the Environment-Economy Interface

There has been a diversity of scientific books and articles describing the development of paradigms and approaches for analyzing the environment-economy interface (e.g. Worster, 1977; Richards, 1986; Braat & van Lierop, 1987; Cleveland, 1987; Martinez-Alier, 1987; Kates, 1988; Turner, 1988; Barbier, 1989; Clark, 1989; Mirowski, 1989; Regier et al., 1989; Underwood & King, 1989).

In their recent book on economics of natural resources and the environment, Pearce and Turner (1990) summarize the evolution of economic paradigms and ideas that have influenced the development of environmental economics (Figure 2). They briefly review the classical, Marxist, neoclassical, and humanistic paradigms, as well as institutional and coevolutionary ways of thinking about natural environments, and emphasize that a pluralistic view of the contribution that economics can make would guard against narrowness in economics, as well as fostering more interdisciplinary analytical linkages, and that this is necessary for improving our understanding of economy-environment interactions.

Colby (1990) has contributed to this understanding by synthesizing many ideas and perspectives from the scientific literature and the environmental debate. He gives his view on the evolution of paradigms in ecology, economics and social systems concerning environmental management in development. He proposes five broad, fundamental paradigms of human-nature relationships, each based on different assumptions about human nature and activity, about nature itself, and about the interactions between nature and humans. Each paradigm asks different questions and perceives different evidence, dominant imperatives, threats or risks, and different preferred solutions and management strategies. Each of them encompasses several schools of thought, and of course there is also some overlapping between the paradigms. Figure 3, modified from Colby (1990), attempts to illustrate the nature of the evolutionary relationships between the paradigms. From *frontier economics* to the diametrically opposite *deep ecology*, paradigms of *environmental protection*, *resource management*, and *ecodevelopment* are evolving. As illustrated in Figure 3, there has been a progression from the two dichotomous paradigms of frontier economics and deep ecology towards perspectives which involves increasing integration of economic, ecological, and social systems. Based on Colby (1990) we will briefly synthesize the major dimensions of these five paradigms.

The type of relationship between society and nature described by the *frontier economics* paradigm is common to both decentralized capitalist economies and centrally-planned Marxist economies, and it has dominated during the industrial development. Although they differ in strategies for organizing development within the economy, the

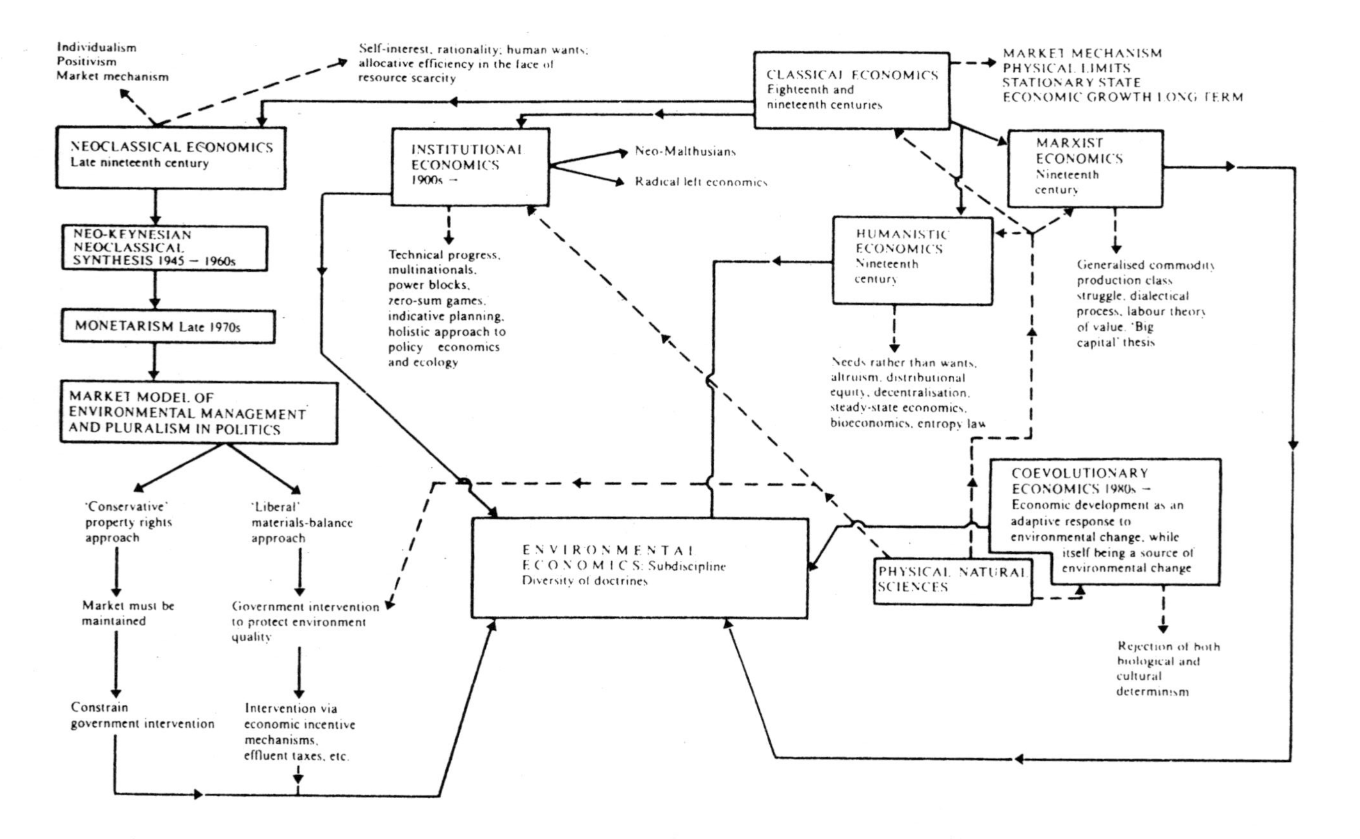

Figure 2. Economic paradigms and the environment. A descriptive model of more or less interconnected theories (from Pearce & Turner, 1990).

underlying world views about humans and nature are similar, often with a vision of infinite economic growth and human progress. From this perspective technologies are developed with the purpose of increasing the power of the socio-economic system to extract resources and increase production of desired goods from the life-supporting environment, as well as to damp the negative impacts of nature's variability on economic activities. It is believed that environmental damage can easily be repaired where necessary, and that infinite technological progress founded in human ingenuity, together with economic growth will provide affordable ways to mitigate environmental problems.

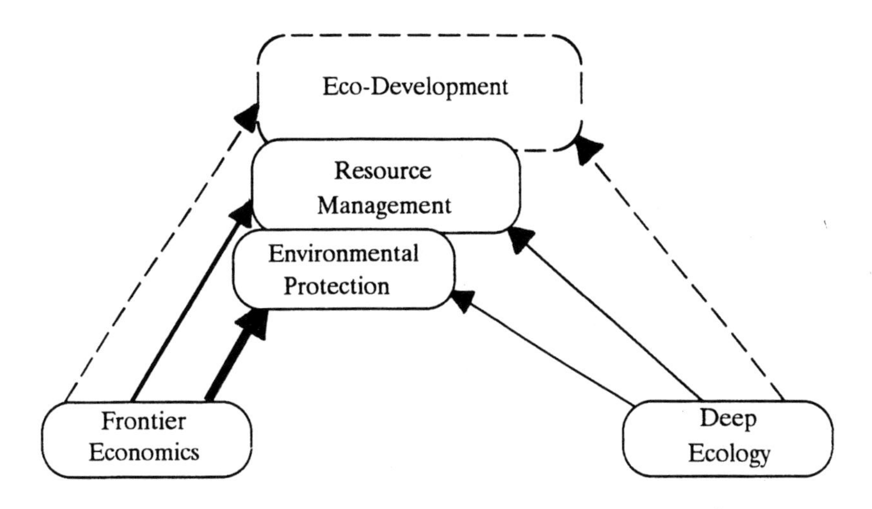

Figure 3. Development of major perspectives in the interface of environment and economy (after Colby, 1990).

The opposite pole is *deep ecology*. This paradigm has not been directly linked to the science of ecology. Is is more of a philosophical, value/ethical movement, generally rejecting the structure and functioning of modern industrial society. It is non-anthropocentric, it emphasizes for example species equality or the intrinsic value of all non-human nature. It stresses the desirability of major reductions in human population, and of bioregional autonomy which means reducing economic, technical, and cultural dependencies and exchanges to confirm them within regions of common ecological characteristics. The deep ecology paradigm also emphasizes promotion of biological and cultural diversity, decentralized planning utilizing multiple value systems, non-growth economies, and simple or low technology levels. The advocates of deep ecology propose major changes in the patterns of human modification of nature, and among deep ecologists there are those who strongly argue for a return to pre-industrial, rural life-styles and standards of living.

The principal strategy of the *environmental protection* paradigm could be expressed as legalizing the environment as an economic externality. It is a modest variation on the frontier economics paradigm of development. The environmental protection perspective is defensive or remedial in practice, concerned mainly with ameliorating the effects of human activities. This approach focuses largely on damage control, on repairing and setting limits to harmful human activities, rather than trying to find ways to improve both development actions and ecological resilience. Governmental agencies are often created and are responsible for setting these limits. Environmental impact statements or assessment are institutionalized in many industrial countries as a rational means of assisting in weighing the costs and benefits of economic development before they are started. Relatively small areas of common property are set aside as state property for preservation or conservation as national parks and wilderness reserves. Resource depletion and ecosystem services are generally not perceived in policy making as serious limiting factors for economic development. The interaction between human activity and nature in the environmental protection paradigm is seen as a question of development versus environment, not recognizing that they are two sides of the same coin.

The basic idea with the *resource management* approach is to incorporate all types of capital and resources (i.e. biophysical, human, infrastructural, and monetary) into calculations of natural accounts, productivity, and policies for development and investment planning. The objective is to take more account of the interdependence and multiple values of various resources, and management of global commons resources are often in focus. It is recognized that ecosystem processes, rather than just stocks of physical resources, need to be considered as resources and capital which should be maintained, as well as used more effectively, by the use of new technology. Resource managers view the stabilization of population levels in developing countries and reductions in the per capita consumption, through increased efficiency, in the industrial nations as essential to achieving sustainability. It is understood that the scale of human activity is now so large that it affects the life-supporting environment as much as Nature affects Man, and that these impacts have a feed-back effect on the quantity and quality of human life that is achievable. The resource management approach is the basic theme of reports such as the Brundtland Report - Our Common Future and the World Resources Institute's annual World Resources reports. The perspective is anthropocentric and the concern for the life-supporting environment is based on the insight that hurting nature is also hurting Man. In a sense ecology is being economized by trying to encompass some basic ecological principles in an attempt to maintain the stability of the life-supporting environment for the support of sustainable development.

Ecodevelopment more explicitly sets out to restructure the relationship between society and nature, by reorganizing human activities so as to be synergetic with ecosystem processes and functions. This emerging paradigm moves from economizing ecology to ecologizing the economy, or whole social systems, and stresses that there are great economic and social benefits to be obtained from fully integrated ecological economic approaches to environmental management. It attempts to move away from the conflict between anthropocentric and biocentric values, but it can be said to follow from the limitations inherent in the environmental and resource management paradigms. The ecodevelopment approach recognizes the need for management of adaptability, resilience, and uncertainty, and for coping with the occurrence of non-linear phenomena and ecological surprises. Rather than asking how can we create, and then, how can we remedy, ecodevelopment attempts to provide a positive, interdependent vision for both human and ecosystem development. This approach emphasizes that planning and management ought to be embedded in the total environment of the system under consideration, including all of the actors concerned, which means that global system awareness must be coupled with local responsibility for action. Eco- signifies both economic and ecological, and the term development, rather than growth, management or protection, connotes an explicit reorientation and upgrading of the level of integration of social, ecological and economic concerns in designing for sustainability. This perspective emphasizes a shift from a system in which the polluter pays to one in which pollution prevention pays, and the need to move from throughput-based physical growth to qualitative improvement. Such development does not only imply becoming more efficient in the use of energy, resources and ecosystem services but it also emphasizes the room for improvements in terms of synergies gained from designing agricultural and industrial processes to mimic and use ecosystem processes in an explicit manner.

As stated by Costanza (1990) "Ecological systems are our best current models of sustainable systems. Better understanding of ecological systems and how they function and maintain themselves can yield insights into designing and managing sustainable economic systems. For example, there is no 'pollution' in climax ecosystems - all waste and by-products are recycled and used somewhere in the system or dissipated. This implies that a characteristic of sustainable economic systems should be a similar 'closing the cycle' by finding economic uses and recycling currently discarded 'pollution', rather than simply storing it, diluting it or changing its state, and allowing it to disrupt existing ecosystems that cannot use it."

According to Colby, the fundamental flaw of Frontier Economics is a lack of awareness of the biophysical basis of human economies (Kåberger, Chapter 4; Hasund, Chapter 7), their dependence on the life-supporting environment (Folke, Chapter 5, 8; Jansson, Chapter 6; Hammer, Chapter 10), and a major criticism of Deep Ecology is that it tends not to be creative, one of the fundamental drives in the evolution of both nature and human society (Bojö, Chapter 3). A major fault with Environmental Protection is that it separates environment and development. A common fault of both Environmental Protection and Resource Management approaches is the mislabeling of various social messes as environmental problems, enabling professionals to conceive them as externalities. Furthermore the myriad problems of development are frequently mismatched with the nature of technical-economic rational logic and its tools on which professionals have come to rely (Andréasson-Gren, Chapter 9; Christiansson, Chapter 11). Colby states that there is a need for a new, mutually positive synthesis of environment-economy development and management, and believes that ecodevelopment (Åshuvud, Chapter 13) is the most promising paradigm for the future. But he also stresses that no single paradigm has the best answer to every type of environmental problem, that change is often resisted due to behavioral and cultural inertia (Svedin, Chapter 1; Söderbaum, Chapter 2), and that there is an urgent need for effective cooperative and institutional innovations to meet the great challenges of the coming decades (Andersson, Chapter 12).

We have used an illustration from Pearce and Turner (1990) (Figure 2 this chapter) to summarize paradigms and ideas that have influenced the development of environmental economics. Colby (1990), based on the work by Herman Daly, classifies the evolution of economic paradigms in terms of allocation, distribution, and scale approaches (Figure 4). He claims that these major concerns of economics, have been seen as separate and conflicting since the late 1800's, with allocative and distributive economics as antagonists in focus, more or less ignoring biophysical issues. Neither free market nor socialist economics or economies have dealt with the necessity of the life-supporting environment and how to maintain and invest in natural capital, a major issue of the evolving ecodevelopment approach. With the risk of being accused of being too imperialistic, we suggest that the recent emergence of the research field of ecological economics could be a new economic synthesis that re-integrates these three types of concern, while at the same time taking into account ecological aspects. Eco-

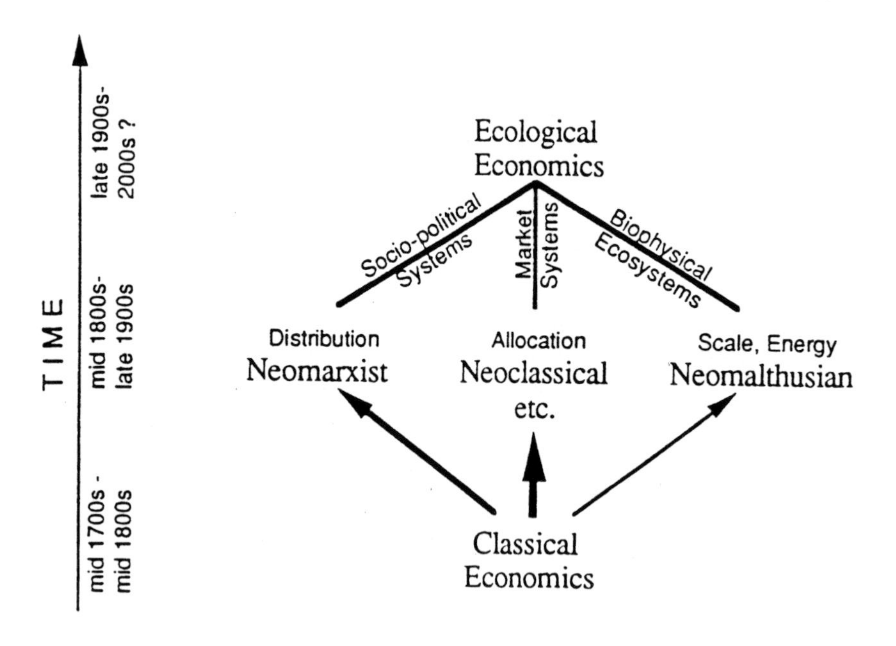

Figure 4. The evolution of economic paradigms from the mid 1700's (after Colby, 1990).

logical economics is concerned with extending and integrating the study and management of ecology and economics, what we might call the ecology of humans and the economy of nature, the web of interconnections uniting the economic subsystem to the global ecosystem of which it is a part. As stated in the first issue of the journal Ecological Economics (Costanza, 1989) this research field is intended to be a new approach to both ecology and economics, that recognizes, the need to make economics more cognizant of ecological impacts and dependencies, the need to make ecology more sensitive to economic forces, incentives, and constraints, and the need to treat integrated economic-ecological systems with a common (but diverse) set of conceptual and analytical tools. Table 2 explicitly discusses some of the major differences between "conventional economics", "conventional ecology" and "ecological economics".

Table 3. Comparison of "Conventional" Economics and Ecology with Ecological Economics (from Costanza, 1991).

	Conventional Economics	**Conventional Ecology**	**Ecological Economics**
Basic World View	Mechanistic, Static, Atomistic individual tastes and preferences taken as given and the dominant force. The resource base viewed as essentially limitless due to technical progress and infinite substitutability	Evolutionary, Atomistic evolution acting at the genetic level viewed as the dominant force. The resource base is limited. Humans are just another species but rarely studied	Dynamic, Systems Evolutionary human preferences evolve to reflect broad ecological opportunities and constraints. Humans are responsible for understanding their role in the larger system and manageing it for sustainability
Time Frame	Short 50 years maximum 1-4 years usual	Multi-Scale days to eons	Multi-Scale days to eons
Space Frame	Local to National country level at best, individual or firm basic unit of analysis	Local to Regional most research focused on relatively small research sites in single-ecosystems	Local to Global hierarchy of scales
Species Frame	Humans Only plants and animals only rarely included for contributary value	Non-Humans Only attempts to find pristine ecosystems untouched by humans	Whole Ecosystems Including Humans acknowledges interconnections between humans and the rest of nature

Table 3, continued

	Conventional Economics	Conventional Ecology	Ecological Economics
Primary Macro Goal	Growth of National Economy	Survival of Species	Sustainability of Whole Planet
Primary Micro Goal	Max Profits (firms) Max Utility (individs) all agents following micro goals leads to macro goal being fulfilled. External costs and benefits given lip service but usually ignored	Max Reproductive Success all agents following micro goals leads to macro goal being fulfilled	Must be Adjusted to Reflect System Goals myoptic following of micro goals can lead to problems which must be compensated for using appropriate cultural institutions
Assumptions About Technical Progress	Very Optimistic	Pessimistic or No Opinion	Prudently Pessimistic
Academic Stance	Disciplinary monistic, focus on mathematical tools	Disciplinary more pluralistic than economics, but still focused on tools and techniques. Few rewards for comprehensive integrative work	Transdisciplinary pluralistic, focus on problems

3. LINKING THE ENVIRONMENT AND THE ECONOMY

3.1 Maintaining Natural Capital

There is a growing concensus, among both ecologists and economists, that maintaining natural capital (see note **1** for a definition), the life-supporting environment on which we depend, is a prerequisite for sustainability. This is illustrated by the following two statements.

1. The necessary condition for sustainable development is *constancy of the natural capital stock*. More strictly, the requirement is for non-negative change in the stock of environmental goods and services (see note **2** for a definition) such as

soil and soil quality, ground and surface water and their quality, land biomass, water biomass, and the waste assimilation capacity of receiving environments (Pearce et al., 1988).

A *working definition* of this statement points to maximizing the net benefits of economic development, subject to maintaining the goods, services and quality of the natural environment over time. This implies that *sustainability can be analyzed in terms of a requirement to maintain the natural capital stock.* This requirement ensures that we observe the *"bounds"* set by the functioning of the natural environment in its role of support system for the economy (Pearce & Turner, 1990).[3]

2. A minimum condition for sustainability is to *maintain the total natural capital stock at or above the current level* (Daly, 1990; Costanza & Daly, 1992).

An *operational definition* of this condition for sustainability means that:
- the human scale must be limited within the carrying capacity of the remaining natural capital
- technological progress should be efficiency-increasing rather than throughput-increasing
- harvesting rates of renewable natural resources should not exceed regeneration rates
- waste emissions should not exceed the assimilative capacity of the environment
- non-renewable resources should be exploited, but at a rate equal to the creation of renewable substitutes.

The significance of maintaining natural capital stems from "the physical dependence of the economic process on its surrounding environment - not just as a source of material and energy inputs but also as an assimilator of waste, and the provider of ecological functions crucial to the maintenance of economic activity and supportive of amenity values, welfare and life in general" (Barbier, 1989).

Pearce and Turner claim that economics has to come to terms with the ecological conditions for sustainability, that *economics need an existence theorem*, a guarantee that any economic optimum, that whatever economy we devise, will not diminish the resilience of ecosystems.

The lack of an existence theorem is connected with a neglect of the scale issue (see Figure 4 and associated text). A well known fact for researchers on the interface between ecology and economics is that the scale and rate of throughput of matter and energy passing through the economic system is subject to an entropy constraint (see Kåberger, Chapter 4; Folke, Chapter 5). Intervention is required because the market by itself is unable to reflect this constraint accurately (Pearce & Turner, 1990). As stated by Daly (1984), "there is no more reason to expect the market to find the optimum scale than there is to expect it to find the optimum income distribution. Just as we impose ethical constraints on income distribution and let market adjust, so must we be willing to impose ecological constraints on the scale of throughput, and let the market adjust." Daly's famous analogy with the plimsoll line clearly illustrates this. Suppose we are economists and want to maximize the load that a boat carries. If we place all the weight in one corner of the boat it will quickly sink or capsize. We need to spread the weight out evenly, and to do this we invent a price system. The higher the waterline in any corner of the boat, the higher the price for putting another kilogram in that corner, and the lower the waterline, the lower the price. This is the internal optimizing rule for allocating space (resources) among weights (alternative uses). This pricing rule is an allocative mechanism only, a very useful but unintelligent computer that sees no reason not to keep on adding weight and distributing it equally until the optimally loaded boat sinks, optimally, to the bottom of the sea. Hence, the price is only a tool for finding the optimal allocation. The optimal scale is something else. What is lacking is a limit on scale,[4] a rule that says "stop when total weight is one ton, or when the waterline reaches the red mark" (Daly, 1984). Simply saying that the end purpose of the economy is to create utility, and to organize the economy accordingly, is to ignore the fact that, ultimately anyway, a closed system such as the Earth sets limits, or boundaries, to what can be done by way of achieving that utility (Pearce & Turner, 1990).

Attempts to estimate where we are in relation to such limits or boundaries have been made by Vitousek, Ehrlich, Ehrlich and Matson (1986) and Wright (1990). They estimated that as much as 20-40 per cent of the global net primary production of natural terrestrial ecosystems is diverted to human activities. The limit on the human scale has also been discussed and estimated for renewable resource sectors, islands and entire regions (e.g. Zucchetto & Jansson, 1985; Reiger & Baskerville, 1986; Folke & Kautsky, 1989; Grima & Berkes, 1989; Odum, 1989; Hammer, Chapter 9; Jansson, Chapter 6; Folke et al., 1991).

In addition to being concerned with growth in overall activity, we also need to consider the substantial negative impacts of current economic activity on the natural environment. Recent data indicate that the productivity of crucial life support ecosystems is now declining in many countries (Clark & Munn, 1986; Brown et al., 1990). We must therefore recognize the environmental costs of economic activities (Bojö, Chapter 3; Hasund, Chapter 7), and incorporate such a recognition in collective and individual behavior (Robinson et al., 1990).

In other words, the multifunctionality of natural capital needs to be acknowledged, including its role as integrated life-support systems. There are risks for the natural capital will be reduced because of our imperfect understanding of the life-support functions, our capability to invent technical substitutes for those functions, generally forgetting that such substitutes require environmental goods and services from other ecosystem - substituting natural capital in one place requires natural capital from somewhere else -, and the fact that losses of life-support functions are often irreversible (Goodland & Ledec, 1987; Daly, 1990; Folke, Chapter 5). In the face of uncertainty and irreversibility, conserving what there is could be a sound risk-aversion strategy (Pearce & Turner, 1990).

3.2. Towards a Sustainable Society

There exists a gallery of definitions on sustainable development, varying from those believing that it is possible to grow in physical terms in a sustainable fashion and to substitute natural capital for man-made capital to those advocating that sustainable development is actually a question of development, a qualitative improvement as opposed to a quantitative growth, emphasizing that there is a strong complementarity between natural capital and the development of human societies (e.g. Odum, 1973; Goodland & Ledec, 1987; Perrings, 1987; Redclift, 1987; Turner, 1988; Daly & Cobb, 1989; James et al., 1989; Pearce et al., 1989; Bojö et al., 1990; Folke, 1990; Folke and Jansson, 1992). There is, however, a consensus that the term *sustainable development* is still vague, and that it means different things for different groups of people involved with environmental-economic issues (Svedin, 1988, Chapter 1 this book).

In addition to the discussion above on natural capital and sustainability we will end this chapter by discussing some principles of sustainability.

Robinson et al. (1990), have attempted to defining a sustainable society. Their principles of sustainability are summarized in Table 2. They stress that in addition to the necessity for the continued existence of biophysical life-support systems - the natural capital - sustainability has socio-political dimensions as well, which implies that a sustainable society must be sustainable in both ecological and socio-political terms. They continue that increasing the sustainability of a system is not equivalent to maintaining the system in its current form, but rather to preserve the capacity for the environmental-socio-political system to change. Sustainable development is continuous development that is never achieved once and for all, but only approached. It is not a state but a process, and as such can only be reinforced not attained.

According to Holling (1986, 1987) the goal is not to increase the resistance to breakdown (the reliability) of the systems, but to increase their capacity to recover from disturbance (their resilience). Bernstein (1981) suggested that natural systems characterized by long-term resilience and persistence can serve as models for the development of an economic system with the same qualities. Sustainable development calls for maintenance of the dynamic capacity to respond adaptively, which is a property of all successful species and societies. Life continues to exist since life has always adapted successfully to and modified the environmental conditions of the planet's surface (Lovelock, 1979, 1988). Thus, it is the vigor and creativity of life that maintains the planet, and our ethical position would be to support this capacity in every way possible. It is directly in our self interest to do so (Golley, 1990). Interpreted this way, sustainable development is a dynamic concept which contrasts with the more conventional idea of development as a series of steps toward a stable state, called maturity and labeled development (Golley, 1990).

Hence, it is not meaningful to measure the absolute sustainability of a society at any point in time. The best that is likely to be possible is to articulate general principles to assess the relative sustainability of the society or the economic activity compared to earlier states or other economic activities. "Our concern should be more with basic

Table 3. Principles of Sustainability (from Robinson et al., 1990)

BASIC VALUE PRINCIPLES

The continued existence of the natural world is inherently good. The natural world and its component life forms, and the ability of the natural world to regenerate itself through its own natural evolution, have intrinsic value.

Cultural sustainability depends on the ability of a society to claim the loyalty of its adherents through the propagation of a set of values that are acceptable to the populace and through the provision of socio-political institutions that make realization of those values possible.

DEFINITION OF SUSTAINABILITY

Sustainability is the persistence over an apparently indefinite future of certain necessary and desired characteristics of the socio-political system and its natural environment.

KEY CHARACTERISTICS OF SUSTAINABILITY

Sustainability is a normative ethical principle. It has both necessary and desirable characteristics. There therefore exists no single version of a sustainable system.

Both environmental/ecological and socio/political sustainability are required for a sustainable society.

We cannot, and do not want to, guarantee persistence of any particular system in perpetuity. We want to preserve the capacity for the system to change. Thus sustainability is never achieved once and for all, but only approached. It is a process, not a state. It will often be easier to identify unsustainability than sustainability.

PRINCIPLES OF ENVIRONMENTAL/ECOLOGICAL SUSTAINABILITY

Life support systems must be protected. This requires decontamination of air, water and soil and reduction in waste flows.

Biotic diversity must be protected and enhanced.

We must maintain or enhance the integrity of ecosystems through careful management of soils and nutrient cycles, and we must develop and implement rehabilitative measures for badly degraded ecosystems.

Preventive and adaptive strategies for responding to the threat of global ecological change are needed.

PRINCIPLES OF SOCIO-POLITICAL SUSTAINABILITY

Derived from environmental/ecological constraints

The physical scale of human activity must be kept below the total carrying capacity of the planetary biosphere.

We must recognize the environmental costs of human activities and develop methods to minimize energy and material use per unit of economic activity, reduce noxious emissions, and permit the decontamination and rehabilitation of degraded ecosystems.

Socio-political and economic equity must be ensured in the transition to a more sustainable society.

Environmental concern needs to be incorporated more directly and extensively into the political decision-making process, through such mechanisms as improved environmental assessment and an environmental bill of rights.

There is a need for increased public involvement in the development, interpretation and implementation of concepts of sustainability.

Political activity must be linked more directly to actual environmental experience through the allocation of political power to more environmentally meaningful jurisdictions, and the promotion of greater local and regional self-reliance.

Derived from socio-political criteria

A sustainable society requires an open, accessible political process that puts effective decision-making power at the level of government closest to the situation and lives of the people affected by a decision.

All persons should have freedom from extreme want and from vulnerability to economic coercion as well as the positive ability to participate creatively and self-directedly in the political and economic system.

There should exist at least a minimum level of equality and social justice, including equality of opportunity to realize one's full human potential, recourse to open and just legal systems, freedom from political repression, access to high quality education, effective access to information, and freedom of religion, speech and assembly.

---

natural and social processes, than with the particular forms those processes take at any time" (Robinson et al., 1990).

Obviously, a better understanding of how to link the natural environment and the economy is urgently required. A dialogue, not a shouting match, is needed between ecologists and economists. For ecologists this will mean not searching for some new economics, but instead understanding better, constructively criticizing but not simply rejecting, what economists have to offer. Economists need to accept the growing importance of maintaining the natural capital, and to renew, discard and adapt theories, concepts and analytical techniques. Economics must also redefine concepts of self-interest to include information about the long-term effects of a decision, and their consequences for the ability to respond adaptively to change (Ehrenfeld, 1978; Bernstein, 1981; Harris, 1985).

Although there are several similarities between the two disciplines, in our view it is necessary both for ecology and economics to broaden current and often too narrowly defined perspectives and paradigms (see Table 1). There is a need for a macroeconomics of the environment combined with a macroecology and integrated with the more traditional micro versions of both ecology and economics (Costanza, 1990). It must be stressed that the macro and micro levels are more complementary than mutually exclusive, both are necessary for improving our understanding of the complex interaction between economic and environmental systems (Barbier, 1989).

We also have to remember that we lose the immense complexities and vast range of qualitative interdependencies that exists in both ecological and economic systems when we analyze the link between the natural environment and the economy unduly in monetary or energy terms (Harris, 1985; Söderbaum, Chapter 2; Kåberger, Chapter 4). It is only possible to understand all the aspects of complex systems by using multiple methodologies (Norgaard, 1989). We definitely also need to study economic-ecological interactions in a context of social responsiveness, of the exercising of economic and political choice, of free will, and of collective action (Andersson, Chapter 12; Svedin, Chapter 1; Söderbaum, Chapter 2), what Berkes and Folke (1992,1993) have termed "cultural capital". Cultural capital refers to all those factors that provide human societies with the means and adaptations to deal with the natural environment and to actively modify it. It includes the wide variety of ways in which societies interact with their life-support environment. Cultural capital will decide how we will use natural capital to "create" man-made capital. Therefore the technologies that we develop can never be value-neutral; they are reflections of our cultural values.

A major condition for moving towards sustainability is an understanding among cultures, that what happens and takes place within our societies not only affects but is also dependent on the life-supporting environment. That the life-supporting environment is the precondition, the basis, for economic development. This basis cannot be replaced by technological substitutes, just as human ingenuity and many technical inventions cannot be replaced by environmental goods and services. Natural capital and man-made capital are complements not substitutes. Man is in his environment as a part of it, and not apart from it. We believe that a very important path towards sustainability is to increase the consciousness of society's dependence on a functioning natural environment. We hope that the essays from the Eco-Eco Group will be a contribution in that direction.

NOTES

1. *Natural capital* has been defined by Costanza and Daly (1992): "(1) renewable or active natural capital and (2) non-renewable or inactive capital. Renewable natural capital is active and self-maintaining using solar energy. Ecosystems are renewable natural capital. They can be harvested to yield ecosystem goods (like wood) but they also yield a flow of ecosystem services" (see note 2). "Non-renewable natural capital is more passive. Fossil fuel and mineral deposits are the best examples. They yield no services until extracted." Natural capital is required to make man-made capital, natural capital includes life-support functions (on which mankind depends) which are not served by man-made capital, natural capital is often multifunctional as opposed to man-made capital, and man-made capital very often generates pollution. Humanly created (man-made) capital refers to capital generated via economic activity.

2. *Environmental goods* refers to both non-renewable resources, such as oil and ore, and renewable natural resources (ecosystem based) such as fish, wood, drinking water and so on. *Environmental services* refers to maintenance of the composition of the atmosphere, of climate, the operation of the hydrological cycle including flood control and drinking water supply, waste assimilation, recycling of nutrients, generation of soils, pollination of crops, provision of food from the sea, maintenance of a vast genetic library and so on (see for example de Groot, 1988, 1992; Ehrlich, 1989; Folke, 1990, Chapter 5). Non-renewable environmental goods are extracted from the life-supporting environment, and renewable environmental goods and environmental services are produced and maintained in the life-supporting ecosystems.

3. A *safe-minimum standard* (Ciriacy-Wantrup, 1968; Bishop & Andersen, 1985) is a non-economic criterion, a safety factor. Safe-minimum standards constrain the economic cost-benefit analysis by specifying environmental, social, or other criteria which the project must meet in all cases.

4. The scale of the economic subsystem is not absolute but relative to the subsystem's "behavior" towards the life-supporting environment.

REFERENCES

Aniansson, B. and Svedin, U. (eds.). 1990. Towards and Ecologically Sustainable Economy. Report from a Policy Seminar. The Swedish Council for Planning and Coordination of Research (FRN), Stockholm.

Barbier, E.B. 1989. Economics, Natural-Resource Scarcity and Development: Conventional and Alternative Views. Earthscan, London.

Berkes, F. and Folke, C. 1992. A Systems Perspective on the Interrelations between Natural, Human-made and Cultural Capital. *Ecological Economics* 5:1-8.

Berkes, F. and Folke, C. 1993. Investing in Cultural Capital for a Sustainable Use of Natural Capital. In: Jansson, A.M., Folke, C., Costanza, R. and Hammer, M. (eds.). Investing in Natural Capital: The Ecological Economic Approach to Sustainability. Island Press, Covelo, California. in press.

Bernstein, B.B. 1981. Ecology and Economics: Complex Systems in Changing Environments. *Annual Review of Ecology and Systematics* 12:309-330.

Bishop, R,C. and Andersen, S.O. (eds.). 1985. Natural Resource Economics: Selected Papers S.V. Ciriacy-Wantrup. Westview Press, London.

Bojö, J., Mäler, K.-G. and Unemo, L. 1990. Environment and Development: An Economic Approach. Kluwer Academic Publishers, Dordrecht.

Braat, L.C. and van Lierop, W.F.J. 1987. Economic-Ecological Modeling. North-Holland/Elsevier, Amsterdam.

Brown, L.R., Durning, A., Flavin, C., French, H., Jacobson, J., Lowe, M., Postel, S., Renner, M., Starke, L. and Young, J. 1990. The State of the World: A Worldwatch Institute Report on Progress Toward a Sustainable Society. Norton, New York. 253 pp.

Ciriacy-Wantrup, S.V. 1968. Resource Conservation: Economics and Policies. University of California, Berkeley, CA.

Clark, M.E. 1989. Ariadne's Thread: The Search for New Modes of Thinking. St. Martin's Press, New York.

Clark, W.C. and Munn, R.E. (eds.). 1986. Sustainable Development of the Biosphere. Cambridge University Press, Cambridge.

Cleveland, C.J. 1987. Biophysical Economics: Historical Perspectives and Current Research Trends. *Ecological Modelling* 38:47-73.

Colby, M.E. 1990. Environmental Management in Development: The Evolution of Paradigms. World Bank Discussion Papers 80. The World Bank, Washington, D.C..

Costanza, R. 1989. What is Ecological Economics? *Ecological Economics* 1:1-7.

Costanza, R. 1990. Ecological Economics as a Framework for Developing Sustainable National Policies. Aniansson, B. and Svedin, S. (eds.). 1990. Towards and Ecologically Sustainable Economy. Report from a Policy Seminar. The Swedish Council for Planning and Coordination of Research (FRN), Stockholm. pp. 45-54.

Costanza, R. (ed.). 1991. Ecological Economics: The Science and Management of Sustainability. Columbia University Press, New York.

Costanza, R. and Daly, H.E. 1992. Natural Capital and Sustainable Development. *Conservation Biology* 6:37-46.

Costanza, R., Daly, H.E. and Bartholomew, J. 1991. Goals, Agenda and Policy Recommendations for Ecological Economics. In: Costanza, R. (ed.). 1991. Ecological Economics: The Science and Management of Sustainability. Columbia University Press, New York. pp. 1-20.

Daly, H.E. 1984. Alternative Strategies for Integrating Economics and Ecology. In: Jansson, A.M. (ed.). Integration of Economy and Ecology: An Outlook for the Eighties. Proceedings from the Wallenberg Symposia. Askö Laboratory, University of Stockholm, Stockholm. pp. 19-29.

Daly, H.E. and Cobb, J.B. 1989. For the Common Good: Redirecting the Economy Toward Community, the Environment and a Sustainable Future. Beacon Press, Boston.

de Groot, R.S. 1988. Environmental Functions: An Analytical Framework for Integrating Environmental and Economic Assessment. Workshop on Integrating Environmental and Economic Assessment: Analytical and Negotiating Approaches, November 17-18, 1988. Canadian Environmental Assessment Research Council, Vancoucer, Canada (mimeographed). 24 pp.

de Groot, R.S. 1992. Functions of Nature. Wolters-Noordhoff, Amsterdam.

Ehrenfeld, D. 1978. The Arrogance of Humanism. Oxford University Press, Oxford.

Ehrlich, P.R. 1989. The Limits to Substitution: Meta-Resource Depletion and a New Economic-Ecological Paradigm. *Ecological Economics* 1:9-16.

Folke, C. 1990. Evaluation of Ecosystem Life-Support in Relation to Salmon and Wetland Exploitation. Ph.D. Dissertation in Ecological Economics. Department of Systems Ecology, Stockholm University, Stockholm.

Folke, C. and Kautsky, N. 1989. The Role of Ecosystems for a Sustainable Development of Aquaculture. *Ambio* 18:234-243.

Folke, C. and Jansson, A.M. 1992. The Emergence of an Ecological Economics Paradigm: Examples from Fisheries and Aquaculture. In: Svedin, U. and Aniansson, B. (eds.). Society and the Environment. Kluwer Academic Publishers, Dordrecht. pp. 69-87.

Folke, C., Hammer, M. and Jansson, A.M. 1991. The Life-Support Value of Ecosystems: A Case Study of the Baltic Sea Region. *Ecological Economics* 3:123-137.

Golley, F.B. 1990. The Ecological Context of a National Policy of Sustainability. In: Aniansson, B. and Svedin, S. (eds.). 1990. Towards and Ecologically Sustainable Economy. Report from a Policy Seminar. The Swedish Council for Planning and Coordination of Research (FRN), Stockholm. pp. 15-25.

Grima, A.P.L. and Berkes, F. 1989. Natural Resources: Access, Right-to-Use and Management. In: Berkes, F. (ed.). Common Property Resources: Ecology and Community-Based Sustainable Development. Belhaven Press, London. pp. 33-54.

Harris, S. 1985. The Economics of Ecology and the Ecology of Economics. *Search* 16:284-290.

Holling, C.S. 1986. The Resilience of Terrestrial Ecosystems: Local Surprise and Global Change. In: Clark, W.C. and Munn, R.E. (eds.). Sustainable Development of the Biosphere. Cambridge University Press, Cambridge. pp. 292-317.

Holling, C.S. 1987. Simplifying the complex: The Paradigms of Ecological function and Structure. *European Journal of Operational Research* 30:139-146.

James, D.E., Nijkamp, P. and Opschoor, J.B. 1989. Ecological Sustainability and Economic Development. In: Archibugi, F. and Nijkamp, P. (eds.). Economy and Ecology: Towards Sustainable Development. Kluwer Academic Publishers, Dordrecht. pp. 27-48.

Kates, R.W. 1988. Theories of Nature, Society and Technology. In: Baark, E. and Svedin, U. (eds.). Man, Nature and Technology: Essays on the Role of Ideological Perceptions. Macmillan Press, London. pp. 7-36.

Lovelock, J. 1979. Gaia: A New Look at Life on Earth. Oxford University Press, Oxford.

Lovelock, J. 1988. The Ages of Gaia: A Biography of Our Living Earth. Oxford University Press, Oxford.

Martinez-Alier, J. 1987. Ecological Economics: Economics, Environment, Society. Basil Blackwell, Oxford.

Mirowski, P. 1989. More Heat than Light: Economics as Social Physics, Physics as Nature's Economics. Cambridge University Press, Cambridge.

Norgaard, R.B. 1989. The Case of Methodological Pluralism. *Ecological Economics* 1:37-57.

Odum, E.P. 1989. Input Management of Production Systems. *Science* 243:177-181.

Odum, H.T. 1973. Energy, Ecology, and Economics. *Ambio* 2:220-227.

Pearce, D.W. and Turner, R.K. 1990. Economics of Natural Resources and the Environment. Harvester/Wheatsheaf, Hemel Hempstead, Hertfordshire, Great Britain.

Pearce, D., Markandya, A. and Barbier, E.B. 1989. Blueprint for a Green Economy. Earthscan, London.

Perrings, C. 1987. Economy and Environment. Cambridge University Press, Cambridge. 179 pp.

Reiger, H.A. and Baskerville, G.L. 1986. Sustainable Redevelopment of Regional Ecosystems Degraded by Exploitive Development. In: Clark, W.C. and Munn, R.E. (eds.). Sustainable Development of the Biosphere, Cambridge University Press, Cambridge. pp. 75-101.

Regier, H.A., Mason, R.V. and Berkes,F. 1989. Reforming the Use of Natural Resources. In: Berkes, F. (ed.). Common Property Resources: Ecology and Community-Based Sustainable Development. Belhaven Press, London. pp. 110-126.

Redclift, M. 1987. Sustainable Development: Exploring the Contradictions. Methuen, London.

Richards, J.F. 1986. World Environmental History and Economic Development. In: Clark, W.C. and Munn, R.E. (eds.). Sustainable Development of the Biosphere. Cambridge University Press, Cambridge. pp. 53-71.

Robinson, J., Francis, G., Legge, R. and Lerner, S. 1990. Defining a Sustainable Society: Values, Principles and Definitions. *Alternatives* 17:36-46.

Svedin, U. 1988. The Concept of Sustainability. In: The Stockholm Group for Studies on Natural Resources Management (eds.). Perspectives of Sustainable Development: Some Critical Issues Related to the Brundtland Report. Stockholm Studies in Natural Resources Management 1. Department of Systems Ecology, Division of Natural Resources Management, Stockholm University, Stockholm. pp. 5-18.

Turner, R.K. 1988. Sustainability, Resource Conservation and Pollution Control: An Overview. In: Turner, R.K. (ed.). Sustainable Environmental Management: Principles and Practice. Belhaven, London. pp. 1-25.

Underwood, D.A. and King, P.G. 1989. On the Ideological Foundations of Environmental Policy. *Ecological Economics* 1:315-334.

Vitousek, P.M., Ehrlich, P.R., Ehrlich, A.H. and Matson, P.A. 1986. Human Appropriation of the Products of Photosynthesis. *BioScience* 36:368-373.

Worster, D. 1977. Nature's Economy: A History of Ecological Ideas. Cambridge University Press, Cambridge.

Wright, D.H. 1990. Human Impacts on Energy Flow through Natural Ecosystems, and Implications for Species Endangerment. *Ambio* 19:189-194.

Zucchetto, J. and Jansson, A.M. 1985. Resources and Society: A Systems Ecology Study of the Island of Gotland, Sweden. Springer Verlag, Heidelberg.

A SELECTION OF RECENT BOOKS IN THE INTERFACE OF ECOLOGY AND ECONOMICS

Ahmad, J.J., Lutz, E. and El Sarafy, S. (eds.). 1989. Environmental Accounting for Sustainable Development. World Bank, Washington, D.C.

Allen, T.F.H. and Hoekstra, T.W. 1992. Toward a Unified Ecology. Columbia University Press, New York.

Archibugi, F. and Nijkamp, P. (eds.). 1989. Economy and Ecology: Towards Sustainable Development. Kluwer Academic Publishers, Dordrecht.

Axelrod, R. 1984. The Evolution of Cooperation. Basic Books, New York.

Baark, E. and Svedin, U. (eds.). 1988. Man, Nature and Technology: Essays on the Role of Ideological Perceptions. Macmillan Press, London.

Barbier, E.B. 1990. Economics, Natural-Resource Scarcity and Development: Conventional and Alternative Views. Earthscan, London.

Berkes, F. (ed.). 1989. Common Property Resources: Ecology of Community-based Sustainable Development. Belhaven Press, London.

Bishop, R.C. and Andersen, S.O. (eds.). 1985. Natural Resource Economics: Selected papers by Ciriacy-Wantrup. Westview Press, New York.

Bojö, J., Mäler, K.-G. and Unemo, L. 1990. Environment and Development: An Economic Approach. Kluwer Academic Publishers, Dordrecht. Second revised and extended revision in 1992.

Bormann, F.H. and Kellert, S.R. (eds.). 1991. Ecology, Economics, Ethics: The Broken Circle. Yale University Press, New Haven.

Botkin, D.B. et al. (eds.). 1989. Changing the Global Environment: Perspectives on Human Involvement. Academic Press, London.

Bowes, M.D. and Krutilla, J.V. 1989. Multiple-Use Management: The Economics of Public Forestlands. Resources for the Future, Washington, D.C.

Braat, L.C. and van Lierop, W.F.J. (eds.). 1987. Economic-Ecological Modeling. North Holland/ Elsevier, Amsterdam.

Brouwer, F. 1987. Integrated Environmental Modelling: Design and Tools. Kluwer Academic Publishers, Dordrecht.

Carter, R.W.G. (eds.). 1988. Coastal Environments: An Introduction to the Physical, Ecological and Cultural Systems of Coastlines. Academic Press, London.

Chatterji, M. & Kuenne, R. (eds.). 1990. Dynamics and Conflict in Regional Structures Change, Macmillan Press, London.

Clark, M.E. 1989. Ariadne's Thread: The Search for New Modes of Thinking. St. Martin's Press, New York.

Clark, W. and Munn, (eds.). 1986. Sustainable Development of the Biosphere. Cambridge University Press, Cambridge.

Collard, D. Pearce, D. Ulph, D. (eds.). 1988. Economics, Growth and Sustainable Environments. Macmillan Press, London.

Conway, R.G. and Barbier, E.B. 1990. After the Green Revolution: Sustainable Agriculture for Development. Earthscan, London.

Costanza, R. (ed.). 1991. Ecological Economics: The Science and Management of Sustainability. Columbia University Press, New York.

Costanza, R., Norton, B.G. and Haskell, B.D. (eds.). 1992. Ecosystem Health; New Goals for Environmental Management. Island Press, Covelo, California.

Daly, H.E. 1991. Steady-State Economics: Second Edition with New Essays. Island Press, Covelo, California.

Daly, H.E. and Cobb, J.B. 1989. For the Common Good: Redirecting the Economy Toward Community, the Environment and a Sustainable Future. Beacon Press, Boston.

de Groot, R.S. 1992. Functions of Nature: Evaluation of Nature in Environmental Planning, Management and Decision Making. Wolters-Noordhoff, Amsterdam.

Dixon, J. and Hufschmidt, M.M. 1986. Economic Valuation Techniques for the Environment. Johns Hopkins University Press, Baltimore, MD.

Dixon, J.A. and Sherman, P.B. 1990. Economics of Protected Areas: A New Look at Costs and Benefits. Island Press, Covelo, California.

Dixon, J.A., James, D.E. and Sherman, P.B. 1989. The Economics of Dryland Management. Earthscan, London.

Dixon, J.A., James, D.E. and Sherman, P.B. (eds.). 1990. Dryland Mangement: Economic Case Studies. Earthscan, London.

Ehrlich, P.R. and Holdren, J.P. (eds.). 1988.The Cassandra Conference: Resources and the Human Predicament. Texas A&M UP, Drawer C, College Station, Texas.

Ehrlich, P.R. and Ehrlich, A.H. 1990. The Population Explosion. Simon and Schuster, New York.

Ellis, D. 1989. Environments at Risk: Case Histories of Impact Assessment. Springer-Verlag, Berlin.

Etnier, C. and Guterstam, B. (eds.). 1991. Ecological Engineering for Wastewater Treatment. Bokskogen, Gothenburg, Sweden.

Faber, M. and Proops, J.L.R. 1990. Evolution, Time, Production and the Environment. Springer-Verlag, Heidelberg.

Faber, M. Niemes, H. and Stephan, G. 1987. Entropy, Environment and Resources. Springer-Verlag, Heidelberg.

Falkenmark, M. and Chapman, T. 1989. An Ecological Approach to Land and Water Resources. Unesco Publications, Paris.

Folmer, H. and van Ierland, E. (eds.). 1989. Valuation Methods and Policy Making in Environmental Economics. Elsevier, Amsterdam.

Freedman, B. 1989. Environmental Ecology: The Impacts of Pollution and Other Stresses on Ecosystem Structure and Function. Academic Press, London.

Friday, L. and Laskey, R. (eds.). 1989. The Fragile Environment. Cambridge University Press, Cambridge.

Funtowicz, S.O. and Ravetz, J.R. 1990. Uncertainty and Quality in Science for Policy. Kluwer Academic Publishers, Dordrecht.

Gever, J., Kaufmann, R., Skole, D. and Vörösmarty, C. 1986. Beyond Oil: The Threat to Food and Fuel in the Coming Decades. Ballinger, Cambridge, Ma.

Goodland, R., Daly, H.E. and El Serafy, S. (eds.). 1992. Population, Technology, and Lifestyle: The Transition to Sustainability. Island Press, Covelo, California.

Goodman, D. and Redclift, M. (eds.). 1989. The International Farm Crisis. Macmillan Press, London.

Grenon, M. and Batisse, M. 1989. Futures of the Mediterranean Basin: The Blue Plan. Oxford University Press/MAB, Oxford.

Hall, C.A.S., Cleveland, C.J. and Kaufmann, R. 1986. Energy and Resource Quality: The Ecology of the Economic Process. John Wiley, New York.

Higashi, M. and Burns, T.P. (eds.). 1991. Theoretical Studeis of Ecosystems: The Network Perspective. Springer-Verlag, Heidelberg.

Hufschmidt, M.M., James, D.E., Meister, A.D., Bower, B.T. and Dixon, J.A. 1983. Environment, Natural Systems, and Development: An Economic Evaluation Guide. Johns Hopkins University Press, Baltimore, MD.

Jansson, A.M. (ed.). 1984. Integration of Economy and Ecology: An Outlook for the Eighties. Department of Systems Ecology, Stockholm University, Stockholm.

Jansson, A.M., Folke, C., Costanza, R. and Hammer, M. (eds.). 1993. Investing in Natural Capital: The Ecological Economic Approach to Sustainability. Island Press, Covelo, California.

Johnston, R.J. 1989. Environmental Problems: Nature, Economy and State. Belhaven Press, London.

Krause, F., Bach, W. and Kooney, J. 1990. Energy Policies in the Greenhouse: From Warming Fate to Warming Limit. Earthscan, London.

Lovelock, J. 1988. The Ages of Gaia: A Biography of Our Living Earth. Oxford University Press, Oxford.

Luken, 1990. Efficiency in Environmental Regulation: A benefit-cost analysis of Alternative Approaches, Kluwer Academic Publishers, Dordrecht.

Marchak, P., Guppy, N. and McMullan, J. (eds.). 1989. Uncommon Property: The Fishing and Fish-Processing Industries in British Columbia. UBC Press, Vancoucer.

Martinez-Alier, J. 1987. Ecological Economics: Energy, Environment and Society. Basil Blackwell, Oxford.

McCay, B.J and Acheson, J.M. (eds.). The Question of the Commons. University of Arizona Press, Tucson, AZ.

McNeely, J.A. 1988. Economics and Biological Diversity: Developing and Using Economic Incentives to Conserve Biological Resources. IUCN, Gland, Switzerland.

Milbrath, L.W. 1989. Envisioning a Sustainable Society: Learning Our Way Out. State University of New York, Albany, NY.

Mitsch, W.J. and Jörgensen, S.-E. (eds.). 1989. Ecological Engineering: An Introduction to Ecotechnology. John Wiley and Sons, New York.

Mirowski, P. 1989. More Heat than Light: Economics as Social Physics, Physics as Nature's Economics. Cambridge University Press, Cambridge.

Mollison, B. Permaculture: A Designer's Manual. Tagari Publ. NSW, Australia.

Myers, N. and Margaris, N.S. (eds.). 1987. Economics of Ecosystem Management. Junk/Kluwer Academic Publishers, Dordrecht

Norton, B.G. (ed.). 1986. The Preservation of Species: The Value of Biological Diversity. Princeton University Press, Princeton, NJ.

Norton, B.G. Toward Unity Among Environmentalists. Oxford University Press, New York.

Odum, E.P. 1989. Ecology and Our Endangered Life-Support Systems. Sinauer, Stamford, CT.

Oldfield, M.L. 1989. The Value of Preserving Genetic Resources. Sinauer, Stamford, CT.

O'Neill,R.V., DeAngelis, D.L., Waide, J.B. and Allen, T.F.H. 1986. A Hierarchical Concept of Ecosystems. Princeton University Press, Princeton.

Orians, G.H., Brown, G.M., Kunin. W.E.and Swierbinski, J.E. (eds.). 1990. The Preservation and Valuation of Biological Resources. University of Washington Press, Seattle.

Ornstein, R. and Ehrlich, P.R. 1989. New World, New Mind: Moving Towards Conscious Evolution. Methuen, London.

Ostrom, E. 1990. Governing the Commons: The Evolution of Institutions for Collective Action. Cambridge University Press, Cambridge.

Parry, M. 1990. Climate change and World Agriculture. Earthscan, London.

Pearce, D. and Turner, K. 1990. Economics of Natural Resources and the Environment. Harvester-Wheatsheaf, Hemel Hempstead, England.

Pearce, D., Markandya, A. and Barbier, E.B. 1989. Blueprint for a Green Economy. Earthscan, London.

Pearson, P. (ed.). 1989. Energy Policies in an Uncertain World. Macmillan Press, London.

Peet, J. 1992. Energy and the Ecological Economics of Sustainability. Island Press, Corvela, California.

Perrings, C. 1987. Economy and Environment. Cambridge University Press, Cambridge.

Pillet, G. and Murota, T. (eds.). 1987. Environmental Economics: The Analysis of a Major Interface. R. Leimgruber, Geneva.

Pimentel, D. and, Hall, C.W. (eds.). 1989. Food and Natural Resources. Academic Press, London.

Pinkerton, E. (ed.). 1989. Co-Operative Management of Local Fisheries. UBC Press, Vancouver.

Pomeroy, L. R. and Alberts, J.J. (eds.). 1988. The Concepts of Ecosystem Ecology. Springer-Verlag, Heidelberg.

Rambler, M.B., Margulis, L. and Fester, R. (eds.). 1988. Global Ecology: Towards a Science of the Biosphere. Academic Press, London.

Redclift, M. 1987. Sustainable Development: Exploring the Contradictions. Methuen, London.

Sagoff, M. 1988. The Economy of the Earth. Cambridge University Press, Cambridge.

Schmidheiny, S. 1992. Changing Course: A Global Business Perspective on Development and Environment. MIT Press, Cambridge, MA.

Siebert, H. 1987. Economics of the Environment: Theory and Policy. Springer-Verlag, Heidelberg.

Smil, V. 1991. General Energetics: Energy in the Biosphere and Civilization. Wiley-Interscience, New York.

Smith, V.K. (ed.). 1988. Enviromental Resources and Applied Welfare Economics: Essays in Honor J.V. Krutilla. Resources for the Future, Washington, DC.

Stockholm Group for Studies on Natural Resources Management (eds.). 1988. Perspectives of Sustainable Development: Some Critical Issues Related to the Brundtland Report. Department of Systems Ecology, Division of Natural Resources Management, Stockholm University, Stockholm, Sweden.

Svedin, U. and Aniansson, B. (eds.). 1992. Society and the Environment. Kluwer Academic Publishers, Dordrecht

Turner, R.K. (ed.). 1988. Sustainable Environmental Management: Principles and Practice. Belhaven/ Westview Press, London.

Turner, R.K. and Jones, T. (eds.). Wetlands: Market and Intervention Failures. Earthscan, London.

Ulanowicz, R.E. 1986. Growth and Development: Ecosystem Phenomenology. Springer-Verlag, Heidelberg.

Western, D. and Pearl, M. (eds.). 1989. Conservation for the Twenty-first Century. Oxford University Press, Oxford.

Wilson, E.O. and Peter, F.M. (eds.). 1988. Biodiversity. National Academy Press/John Wiley and Sons, New York.

Wulff, F., Mann, K.H. and Field, J.G. (eds.). Network Analysis in Marine Ecology. Springer-Verlag, Heidelberg.

Zucchetto, J. and Jansson, A.M. 1985. Resources and Society: A Systems Ecology Study of the Island of Gotland, Sweden. Springer Verlag, Heidelberg.

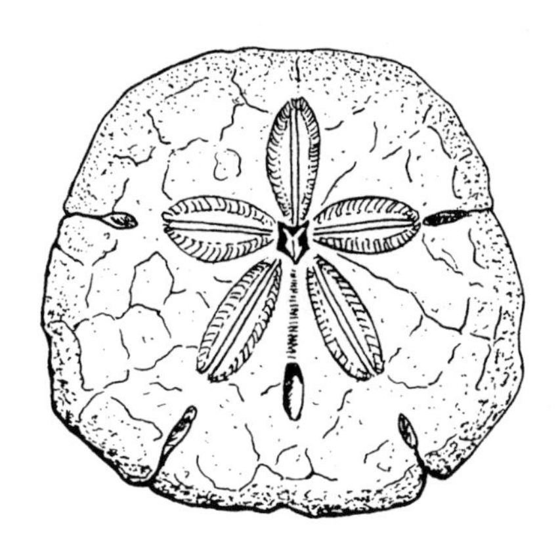

The Eco-Eco Group

The initiative to the Eco-Eco Group was taken by the late Johan Åshuvud, to whom this book is dedicated. Johan worked with enthusiasm to gather young scientists from various disciplines, and with the help of Ann Mari Jansson and Carl Folke the Eco-Eco Group was founded in 1984. Back in 1984 the research in the interface of ecology and economy had just started to spread worldwide. The first international conference on these matters "Integration of Economy and Ecology: An Outlook for the Eighties" (Jansson, 1984), took place in Stockholm, in 1982. Several of the Swedish participants in that conference actively work in this interdisciplinary research field, and are also members of the Eco-Eco Group.

The Group consists of scientists and Ph.D. students with a diverse background, such as systems ecology, marine ecology, botany, hydrology, geography, forestry, agriculture, physical geography, physics, natural resource, environmental and agricultural economics, business and administration, economic geography, economic history, international economics, engineering, regional planning, and political sciences. The interdisciplinary mix challenges the participants to broaden their perspective, which is necessary when working with complex systems and issues in the interface of linking the natural environment and the economy.

The major purpose of the Eco-Eco Group is to stimulate scientific communication and cooperation between scientists from various disciplines, interested in environmental and sustainability issues.

Topics often discussed are for example, valuations of natural resources, environments and ecosystems, environment and development in developed as well as in developing countries, carrying capacity issues, limits to the stresses on ecosystems, nature conservation, ecological engineering, regional balance issues, relation between sectors, the role of infrastructures, the transition from a throughput based society to a sustainable society, basic assumptions underlying current economic and ecological paradigms and the implications of alternative assumptions, the different use and meaning of similar terms and concepts in ecology and economics, thermodynamics for economics and ecology, environmental ethics, ecological-economic modelling, methods of implementing efficient environmental policies, economic incentives and policy instruments.

The meetings are held in a very informal way, thereby enhancing constructive discussions and the generation of new ideas, while at the same time avoiding intradisciplinary "rules of the game." The meetings generally consist of integrated presentations-discussions by the members of the group or by invited scientists.

Although there is a variety of different positions within the group, as clearly reflected by the essays of this book, we believe that the members have taken the challenge of broadening their perspectives. This is beneficial and of great significance for the development of integrated ecological economic theories, tools, and policies aiming at approaching sustainable environmental management and socio-economic development.

REFERENCES

Jansson, A.M. (ed.). 1984. Integration of Economy and Ecology: An Outlook for the Eighties. Proceedings from the Wallenberg Symposia. Askö Laboratory, Stockholm University, Stockholm.

The logo of the Eco-Eco Group is the sand-dollar, a symbol of the cooperation between ecologists and economists.

NOTES ON THE CONTRIBUTORS

Thomas Andersson, is Director of the International Research Programme at the Industrial Institute for Social and Economic Research (IUI). He is also affiliated with Stockholm School of Economics, from which he has a Ph.D. He has been a Visiting Scholar at the Bank of Japan and a Visiting Fellow at, e.g. Harvard University, Boston, Universidade de Sao Paulo, and the EastWest Center, Honolulu. His research concerns foreign direct investment, the relations between Europe and East Asia, and international environmental issues.

Jan Bojö, Ph.D. is an environmental economist at the World Bank, Washington, D.C. His previous research at the Stockholm School of Economics focused on the economics of land degradation and rehabilitation.

Carl Christiansson, holds a Ph.D. in Physical Geography. He is Associate Professor in Natural Resources Research and head of the Environment and Development Studies Unit at the School of Geography, Stockholm University. Between 1987-1990 he was research leader at the Department of Ecology and Environmental Research, Swedish University of Agricultural Sciences. He is presently coordinator of cooperation programmes on research and higher education financed by the Swedish Agency for Research Cooperation with Developing Countries (SAREC). He has been involved in environmental research in Tanzania and Kenya since 1970, and is presently project leader of SAREC and SIDA (the Swedish International Development Authority) financed studies in Botswana and Tanzania, and advisor to SIDA and the Swedish Society for Nature Conservation on tropical land management issues. He has extensive teaching experience in matters concerning tropical environments, has conducted field courses in tropical geography between 1977-1988. He has published some 60 scientific papers, reports and articles mainly on tropical land use, soil erosion and conservation, environmental change and socio-economics of rural Africa.

Carl Folke, Ph.D., is Deputy Director of the Beijer International Institute of Ecological Economics, the Royal Swedish Academy of Sciences, and part time researcher at the Department of Systems Ecology, Stockholm University. His research is on the role and value of ecosystems as life-support systems to human societies. Together with Ann Mari Jansson and Monica Hammer, he organized the Second Meeting of the International Society for Ecological Economics - Investing in Natural Capital: A Prerequisite for Sustainability - in Stockholm, 1992. He is the Book Review Editor of the journal Ecological Economics, and serves as advisor to the Swedish Minister of the Environment and Natural Resources.

Ing-Marie Gren, Econ.Dr., is researcher at the Beijer International Institute of Ecological Economics, the Royal Swedish Academy of Sciences, and partly at Department of Economics, Swedish University of Agricultural Sciences. She has been involved in interdisciplinary research on regional environmental management for a decade. She is currently engaged in a European joint research programme on environmental issues, and is a review editor of the journal Environmental and Resource Economics.

Monica Hammer, holds a B.Sc. in economics and busines, administration as well as in biology with emphasis on ecology. Her research, as a Ph.D. student, in Natural Resources Management at the Department of Systems Ecology, concerns the interface between ecology and economy with special reference to the Baltic fisheries. She is presently engaged in a project concerning integrated ecological economic aspects of Baltic Fisheries and environmental changes, ecosystem support and common property issues.

Knut Per Hasund has a background in economics as well as biology (M.Sc. in agricultural engineering) and holds a Licentiate in Agronomy. His research deals with resource and environmental economics within the agricultural sector, now studying recreational, environmental, cultural-historical and agricultural values of cultivated land and landscapes, and related policy instrument analysis. He is employed as Assistant Professor at the Department of Economics, the Swedish University of Agricultural Sciences.

Ann Mari Jansson, Professor at the Department of Systems Ecology, Stockholm University, is director of the Center for Natural Resources and the Environment, Stockholm University, and a co-editor of the international journal Ecological Economics. She has among other things been Swedish delegate of Unesco since 1980 and chairman of the Science Committé of the Swedish Unesco Council. Besides various scientific appointments she has served as an expert to the State Department of Agriculture and been a member of the Scientific Advisory Board to the Swedish Government. She has also been chief examiner of the Environmental Systems Teaching Program, at the International Baccalaureate School. Her scientific papers involve ecological and economic interrelations, systems analysis of various subsystems, especially in the Baltic drainage basin including a regional ecological/economic study of the island of Gotland, Sweden.

Tomas Kåberger holds a M.Sc. in Engineering Physics and has also studied Human Ecology, Eco-philosophy, and Economics. He has a Licentiate of Engineering degree in Physical Resource Theory from Chalmers University of Technology, where he has been lecturing Natural Resources in the Economy. Besides his work at the university, he has worked for the Swedish International Development Authority and the Swedish Society for Nature Conservation. He is presently working part time for Ecotraffic AB.

Uno Svedin, Ph.D., is executive secretary of the Committee for Natural Resources, Swedish Council for Planning and Coordination of Research. He is also Associate Professor of Physics, especially Systems Analysis, at the Department of Physics, Stockholm University. He has been responsible for the Swedish national report to the UN Conference on Science and Technology for Development (UNCSTD) 1979 and been Swedish representative to various fora in international organizations such as UNDO, OECD, UNESCO. Publications involve topics as natural resources management, cultural aspects of man's use of Nature, substitution processes, methods of future studies.

Peter Söderbaum is Associate Professor of natural resource management at the Department of Economics, Swedish University of Agricultural Sciences. He received his Ph.D. from the Department of Business Administration, Uppsala University and is International Correspondent for the Association for Evolutionary Economics with the Journal of Economic Issues. His interests include institutional economics and other non-conventional perspectives, approaches to decision-making, and actor's perspective as an alternative to public choice theory.